Bioinformatics for agriculture: High-throughput approaches

Atul Kumar Upadhyay • R Sowdhamini •
Virupaksh U. Patil
Editors

Bioinformatics for agriculture: High-throughput approaches

Springer

Editors
Atul Kumar Upadhyay
Department of Biotechnology
Thapar University
Patiala, Punjab, India

R Sowdhamini
Department of Biochemistry, Biophysics and Bioinformatics
National Centre for Biological Sciences
Bangalore, India

Virupaksh U. Patil
Division of Crop Improvement
Central Potato Research Institute
Shimla, Himachal Pradesh, India

ISBN 978-981-33-4790-8 ISBN 978-981-33-4791-5 (eBook)
https://doi.org/10.1007/978-981-33-4791-5

This Springer imprint is published by the registered company Springer Nature Singapore Pte Ltd.
The registered company address is: 152 Beach Road, #21-01/04 Gateway East, Singapore 189721, Singapore

Contents

Introduction to the Concepts of Agr-Informatics

Sidharth Singh, M. Aafikul Haque, and Om Silakari

Introduction

What is Bioinformatics? Defining a term like Bioinformatics is a tedious task especially because it essentially has multiple meanings covering a wide range of topics which includes but not limited to DNA data storage, mathematical modeling, understanding the mechanism behind complicated human diseases, etc. In basic terms bioinformatics can be defined as an interdisciplinary research area which amalgamates biological and computational science [1]. How? In the modern world, computers are being used by biologists equally as compared to the individuals of any other profession; for example, Bankers or pilots. Apart from sending regular emails, filling up spreadsheets, listening to music, biologists are trained to do certain specific tasks like storing complex biological information into various databases, developing algorithms to retrieve meaningful data from these databases, designing mathematical models to predict the outcome, etc. [2]. In our personal understanding we can say that Bioinformatics is the science which deals with the computational management of all kinds of molecular biological information. Now the question arises what is molecular biological information and how computational science is managing it? To understand this, we have to look at a wider picture on the biological work which is being done in the world. At present most labs are generating data which is related to the mechanism of a particular disease, new finding in already established concepts, developing new drugs, new vaccines, etc. So, we need a place

S. Singh (✉)
Department of Biotechnology, Thapar Institute of Engineering & Technology, Patiala, India

M. A. Haque
Department of Pharmaceutical Analysis, Anurag University, Hyderabad, India

O. Silakari
Department of Pharmaceutical Science and Drug Research, Punjabi University, Patiala, Punjab, India

A. K. Upadhyay et al. (eds.), *Bioinformatics for agriculture: High-throughput approaches*, https://doi.org/10.1007/978-981-33-4791-5_1

where all this data can be stored and retrieved when needed by anyone throughout the world. This aspect is taken care of by computational science.

At times the line which can distinguish biology from computational science is very blurry, hence many people use Bioinformatics and computational biology terms invariably [3]. In our view Bioinformatics and computational biology are interdisciplinary fields which involve researchers having expertise in different fields; for example, computer science, molecular biology, genetics, mathematics, statistics, physics, etc. Aim of these two fields can be defined as follows:

a) Bioinformatics: It is concerned with collecting and storing the biological information. Anything related to biological databases are included in this field.
b) Computational biology: It is concerned with the development of computational programs/algorithms and various statistical models needed to understand the complex biological data

We can further understand these two fields according to the definition given by NIH.

Bioinformatics: Research, development, or application of computational tools and approaches for expanding the use of biological, medical, behavioral, or health data, including those to acquire, store, organize, archive, analyze, or visualize such data [1].

Computational Biology: The development and application of data-analytical and theoretical methods, mathematical modeling and computational simulation techniques to the study of biological, behavioral, and social systems [1].

So why has Bioinformatics become the new buzz word in the market?

Human genome project was a 13-year-long international, collaborative program which aims to determine all the genes that make up the human genome [4]. Once it was completed almost half of the genes identified have no known function. Bioinformatics has played a key role in identifying, establishing their functions and further understanding their role in diagnosing, preventing, and treating diseases. Bioinformatics can guide us to uncover hidden information in our DNA and moreover it can help companies to save money and time [5]. In addition to this a lot of data is being generated due to advances in biotechnology techniques which need to be managed. The old rule of supply and demand is also favoring the rise of bioinformatics. There are not many people who are adequately trained in both fields, i.e. biology and computer science. So, we need a biologist who can work with computers or vice versa to solve the modern day biotechnology problems. We can conclude that there are four main reasons which have played a significant role in rise of bioinformatics:

a) First, is the never-ending collection of DNA and protein sequences which we are generating at a pace faster than ever, this data needs to be managed so that it can be used to solve the problems which earlier were impossible to solve.
b) Second, to generate a meaningful conclusion from the raw data (DNA and protein sequences) computer algorithms, programs as well as number crunching power of computers are needed.

c) Third, is the availability of high-powered computers to biologist which till World War II were not available.
d) Fourth, is the idea that macromolecules are the source of a variety of information becomes the core of scientific discoveries and is being considered as the connecting link between computational biology and molecular biology which leads to acceptance of computer science by biologists.

We have now understood why there is a buzz for Bioinformatics in the market; but another question remains unanswered, what the aims of Bioinformatics are? Or what we can achieve with the use of Bioinformatics?

a) First and foremost is the organization of existing data so that researchers can easily access the information, they must also be able to submit new entries as and when they are generated, e.g. DDBJ for DNA related information.
b) All the information which is stored in various databases are essentially useless until they can be analyzed. So, another goal of bioinformatics is to develop tool and resources to analyze data [6].
c) Third aim is to successfully implement these tools to analyze data and draw logical conclusion [7].

Some Basic Biological Science

In order to understand, implement, and improve the existing bioinformatics tools and techniques for smooth analysis of raw data, one must first be well acquainted with basics of biological science especially molecular biology.

Scale and Time

Biology can be defined as the science of life and all living organism present on this earth. An organism is a living entity composed of a single cell, e.g. bacteria, or multiple cells, e.g. animals, plants, etc. Multicellular organisms are visible to human eyes but single cell organism cannot be visualized by naked human eye owing to their very small size, ranging between 1 μm to 100 nm (some viruses) [8]. All the life forms are in essence made up of small molecules having a size range of about 1 nm. Because it is difficult to visualize anything at this small scale so scientists have to design novel visualization techniques to see these molecules. These techniques have generated huge amount of data due to which scientists were able to understand the complex nature of various human life processes.

Life on earth began long back almost 4 billion years ago, not very long after earth comes in to existence. Since then, all life forms have seen evolution happening during the course of their life. If one has to compress earth's evolution in terms of time, let us say in a month, then we can say that origin of life happened around first 3–4 days but majority of the life forms came into existence only after the 27th day

[9]. A lot of multicellular organism appeared in the last few days, i.e. land plants and land animals came in to existence on the 28th day, mammals came in to existence on 29th day, and on the last day all the birds and flowering plants appeared on earth. Modern day's humans or homo sapiens came in to existence in the last 10 mins of the last day. The process of gradual change in the genetic composition of a living organism and becoming more complex to adapt to new changing environment is known as evolution [10]. The process of evolution often ends with the development of a new species. While studying evolution, it is important to realize that the tree of evolution is composed of various branches or leaves each depicting a species which has evolved from the previous one. So, in order to completely understand evolution of a particular specie one must be able to compare related species.

Cell

Cell can be defined as the smallest structural and functional unit of an organism. Some organisms are single cellular, i.e. they have only one cell, whereas some other are multicellular having billions of cells, e.g. plants and animals.

Broadly cells can be divided in to two types: Prokaryotic and eukaryotic cells. The major difference between prokaryotes and eukaryotes is the absence of nucleus in the former and presence in the later [11]. Living organism are also divided in to prokaryotes and eukaryotes based on the presence of prokaryotic cell and eukaryotic cells. The earliest form of life on earth mainly composed of prokaryotes which include single cell organism like bacteria and archaea. Eukaryotes mainly composed of higher animals like plants and animals and certain unicellular organism like yeast. *E. coli*, a bacterium, is the most extensively studied prokaryote having a very simple life process; on the other hand, eukaryotic cells consist of more complex structures commonly known as organelles [11]. The nucleus of the eukaryotic cells contains the genetic material DNA which is coiled or compressed in to chromatin or chromosomes. When a cell is not participating or undergoing cell division the proteins and DNA are aggregated to form chromatin which is scattered all over the nucleus [12]. When cell starts dividing these chromatin molecules further gets packed in to a structure commonly known as chromosomes. It has two arms the P-arm (shorter arm) and Q arm (longer arm) joined to each other with the help of centromere.

DNA and Chromosome

DNA stands for deoxyribonucleic acid; it stores all the genetic information of a cell. Some organism has RNA as the genetic material like some virus (Coronavirus). DNA and RNA are made up of nucleotides. Nucleotides itself are composed of bases (adenine, thymine, guanine, and cytosine), a molecule of pentose sugar and phosphoric acid [13]. Nucleosides are composed of a molecule of pentose sugar and a nitrogenous base as shown in Fig. 1. The sugar molecule present in DNA molecule is

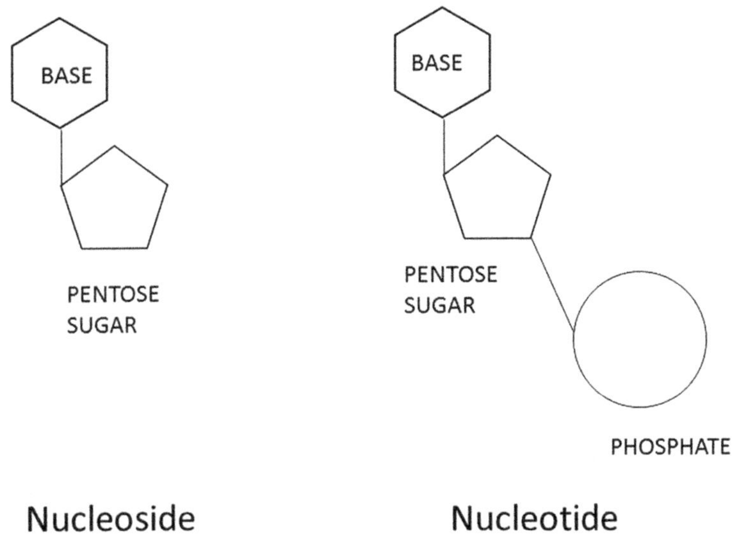

Fig. 1 Composition of nucleoside and nucleotide

de-oxy ribose sugar, while ribose sugar in present in an RNA molecule. Bases can be divided in to two groups: purine and pyrimidine—adenine (A) and guanine (G) are purines with two fused rings, while thymidine (T) and cytosine (C) having single ring belong to the category of pyrimidine [13].

Thymidine (T) is replaced by a new base Uracil (U) in a RNA molecule. DNA is composed of two complementary strands which run in opposite direction, i.e. one is running in 5′ to 3′ direction and another one in 3′ to 5′ direction. Pentose sugar and phosphate group act as backbone of the entire DNA/RNA molecule.

The famous double helix of DNA is made up of two complementary strands having hydrogen bonds between them (A and T have two hydrogen bonds, while G and C have three hydrogen bonds). This bonding is named as base pairing. Generally, RNA is present in nature as single stranded molecule but occasionally it pairs with DNA, the base pairing rule for this situation is A-U, T-A, G-C, and C-G [14].

The pentose sugar present in the DNA/RNA molecule has five carbon atoms numbered as 1′–5′. DNA or RNA molecule generally referred to as running in a 5′ to 3′ direction or 5′ end/3′end, this naming convention is also based on pentose sugar numbering system. The DNA sequence is always read in 5′ to 3′ direction.

5′-A T T A C G G T A C C G T -3′
3′-T A A T G C C A T G G C A -5′

DNA is a very complex molecule and it is present inside the nucleus in a highly compact form. It binds with histones to form nucleosomes that appear as beads on a DNA string. This nucleosome gets compressed by coiling again and again to form supercoiled chromatin fibers. This coiled structure further folds to form loops and on further coiling chromosomes are formed. The entire length of a DNA molecule inside a single cell is approximately 2 m but due to this supercoiling phenomenon, the entire DNA molecule can fit inside a nucleus of diameter approximately 5 μm [15]

Central Dogma

The central dogma in biology is described as the mechanism through which the encoded information in DNA is passed on to messenger RNA (mRNA), and then further used to direct the synthesis of proteins. The former process is known as transcription (DNA→mRNA), while the latter is known as translation (mRNA→proteins). Proteins are made up of small chains of amino acids joined together by peptide bond [16]. There are 20 different types of standard amino acid. The synthesis of proteins is based on the mRNA sequence which is tightly regulated by a universal genetic code as depicted in Fig. 2. A genetic code is a three-letter code made up of nucleotides, is specific for each amino acid, and is referred to as triplet. Since 64 unique amino acids can be coded by three nucleotides the presence of redundancies is inevitable [18]. One codon can code for more than one amino acid.

	U	C	A	G	
U	UUU Phenyl UUC alanine UUA Leucine UUG	UCU UCC Serine UCA UCG	UAU Tyrosine UAC UAA Stop UAG codon	UGU Cysteine UGC UGA Stop codon UGG Tyrptophan	U C A G
C	CUU CUC Leucine CUA CUG	CCU CCC Proline CCA CCG	CAU Histidine CAC CAA Glutamine CAG	CGU CGC Arginine CGA CGG	U C A G
A	AUU AUC Isoleucine AUA AUG Methionine; Start codon	ACU ACC Threonine ACA ACG	AAU Asparagine AAC AAA Lysine AAG	AGU Serine AGC AGA Arginine AGG	U C A G
G	GUU GUC Valine GUA GUG	GCU GCC Alanine GCA GCG	GAU Aspartic GAC acid GAA Glutamic GAG acid	GGU GGC Glycine GGA GGG	U C A G

Fig. 2 Standard genetic code

All the redundant codons have first two identical nucleotides differing only in the last nucleotide. The start codon for any protein molecule is AUG which codes for methionine, while there are three different stop codons, viz, CAA, CAG, UGA.

The process of protein synthesis in prokaryotes is different from the process in eukaryotes. In prokaryotes, after the unwinding of DNA, one strand is used as a template to generate mRNA which is simultaneously used to synthesize proteins with the help of tRNA [17]. In eukaryotes, first half takes place inside the nucleus where DNA resides and mRNA is synthesized. Synthesized mRNA is referred to as pre-mRNA. The second half takes places in the cytoplasm so pre-mRNA needs to undergo certain modification like a poly A tails is added to protect it from the cytoplasmic enzymes and certain parts are removed referred to as splicing [18]. After entering the cytoplasm the process of translation begins and with the help of tRNA protein molecules are synthesized.

Till here we have talked about the basic biological science which one must know before diving in to the ocean of bioinformatics as this knowledge helps one to understand when, what, and how to apply bioinformatics to get desired results. From here onwards we will be discussing various bioinformatics tools, databases, online resources which are frequently used by a modern day bioinformatician.

Databases

Due to the advent of new technologies, huge amount of raw sequence data is being generated nowadays and as the volume of this data grows, modern day bioinformatician needs to develop more sophisticated computational tools to manage this data. Hence management of this huge volume of information becomes imperative, so there is need to constantly improve and develop new, advanced computer databases. A Biological database is no different from a conventional computer database except it stores complex biological data, e.g. DNA/protein sequences [19]. Like all databases it is organized, managed by computational algorithms which help in constant submission, retrieval, and updating of the database. In simple terms a database can be defined as an organized collection of raw data, generally stored and accessed with the help of a computer to generate meaningful results. Example: DDBJ (DNA Data Bank of Japan) which collects DNA sequences, PDB (Protein Data Bank) which stores structures of proteins.

Types of Biological Databases

Biological databases can be broadly divided in two categories: On the basis of source and on the basis of nature of data. On the basis of source there are two types of databases: primary databases and secondary databases [20].

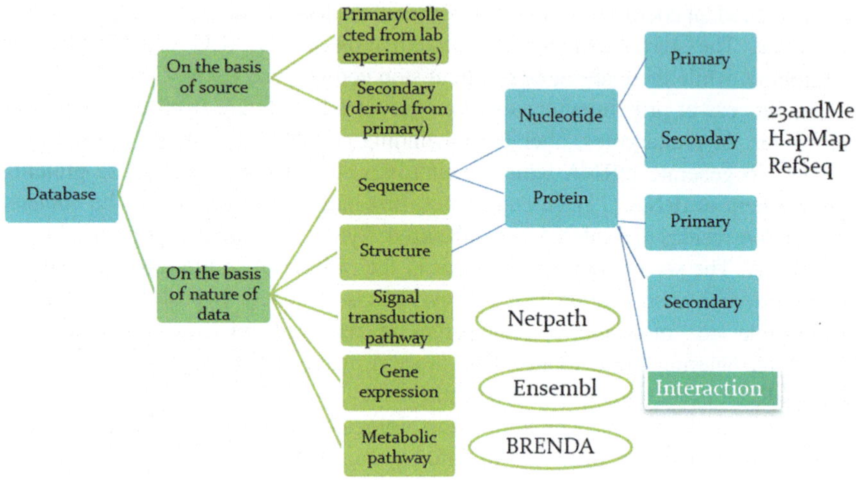

Fig. 3 Different types of databases

On the basis of nature of data, databases can be broadly divided in to 5 sub categories: Sequence database, structure database, signal transduction pathway database, gene expression database, and metabolic pathway database as shown in Fig. 3.

Primary Databases

Primary databases contain experimentally generated raw data such as DNA and protein sequences. Scientist working all over the world directly submits their experimental data after which an accession number is given to each entry so that it can be easily retrieved wherever necessary [21].

Example: DDBJ and Genbank for genome sequences, Swiss-Prot and PIR for protein sequences, Protein Data Bank for three-dimensional protein structures.

Secondary Databases

Secondary databases store information which is derived from the primary data. They generally store information derived from various resources like scientific literature and other databases [21]. They are highly curated and one can find information related to conserved sequences, active site residue, and signature sequences in these databases. Examples: SCOP, CATH, PROSITE.

Special Databases

Apart from all these databases there are certain special databases which cater to a specialized research interest, e.g. OMIM inherited diseases database (Online Mendelian Inheritance in Man), Gene expression omnibus—Microarray database, Array expression database—Microarray database, Whole genome database—ENSEMBL.

Sequence Alignment

What is a sequence alignment? What is the purpose of aligning two sequences? How to do sequence alignment? These all are the question which comes to mind as soon as someone talks about sequence alignment. In simple terms sequence alignment is a technique/process which tells us that how much similarity is present between two or more sequences (the sequence can be of DNA, RNA, or proteins), is there any evolutionary relationship between the sequences? [23]. In an alignment of two sequences (protein or nucleotide), certain parts of the sequence will be exactly matching, we call them as "match," while some other parts will not match, we call them as "mismatch" [22]. Let us understand sequence alignment with the help of an example:

Consider two sequences **abcdef** and **abdgf,** we have to align them. Write second sequence below the first one

abcdef

abdgf

Now move sequence in order to generate a match between two sequences.

abcdef

| |

abdgf

The characters that are matching are marked with vertical lines. In order to *maximize the alignment*, we inserted a gap between **b** and **d** in the lower sequence.

abcdef

|| | |

ab-dgf

In this alignment e and g do not match. So, the goal of aligning two sequences is

- To maximize base to base matches.
- If required insert gaps in either of the sequence so that the overall alignment can be made.
- The order of bases in each sequence must remain preserved.
- Gap to gap match is not considered.

Now how to evaluate if the alignment generated by the above method is a good alignment? For this we need some sort of scoring scheme. A scoring scheme consists of score for each possible replacement of bases (positive score for a match and negative score for a mismatch) and penalties are added when gaps are encountered. An overall alignment score is generated by adding all the substitution score and gap penalties. The higher the score, better the alignment. In the above example if we assign the following scoring pattern: Match = +1, Mismatch = −1, and Gap = 0, then we will get final score as 3 (1 + 1 + 0 + 1 − 1 + 1).

Let us consider another example to calculate which the best alignment is.

Query seq:	ATGGCG
Seq1:	ATGAG

Now we can align Seq1 in two below mentioned ways:

Query:	ATGGCG
Alignment 1: ATG_AG	Score +1 + 1 + 1 + 0 − 1 + 1 = 3
Alignment 2: A_TGAG	Score +1 + 0 − 1 + 1 − 1 + 1 = 1

As discussed earlier the higher the score, the better is the alignment. Hence, we consider alignment 1 as the best alignment.

What Is the Need of Sequence Alignment?

If two protein or nucleotide sequences are aligned, then the degree of similarity between two sequences can tell us how closely related they are; in other words do they have a common ancestor or not? Since the secondary, tertiary, and quaternary structures of proteins are dependent on the sequence of protein so if two proteins share similarity in the sequence, then we can say, they share similarity in the structure and function. Hence the purpose of doing sequence alignment is to determine the functionality of unknown protein or to find conserved region within a nucleotide sequence [23].

Alignment Methods

The sequences which are very similar or very short can be aligned manually. However, in most cases the alignment is done between very lengthy and immensely variable sequences which are very difficult to align manually. Hence a number of algorithms and computer programs have been designed to make our work easier. With the help of these algorithms very complex, highly variable, and immensely lengthy sequences can be aligned in a matter of few minutes. These tools also help us to interpret the results by strategically presenting the patterns in the results which are otherwise difficult to show algorithmically [24]. Computational methodologies used for sequence alignment can be divided in to two types: pair wise sequence alignment and multiple sequence alignment. The major difference between these two is the former is used only when an alignment needs to be generated between two sequences, whereas the latter is used to identify similar region (which may indicate functional, structural, and evolutionary connection) among three or more sequences [25]. Further, pair wise sequence alignment is divided in to local and global alignment (Fig. 4). In global alignment an end to end alignment of two sequences is created to check the similarity between two sequences, while in local alignment the algorithm looks for one or more small areas that show similarity within the two sequences. A number of computational programs have been used for aligning sequences. These include dynamic programming which is slow, and probabilistic method which is used for large-scale database search but it does not guarantee to give the best results. For generating global alignment Needleman–Wunsch algorithm [26] is used, whereas Smith–Waterman algorithm [27] is used to generate local alignment.

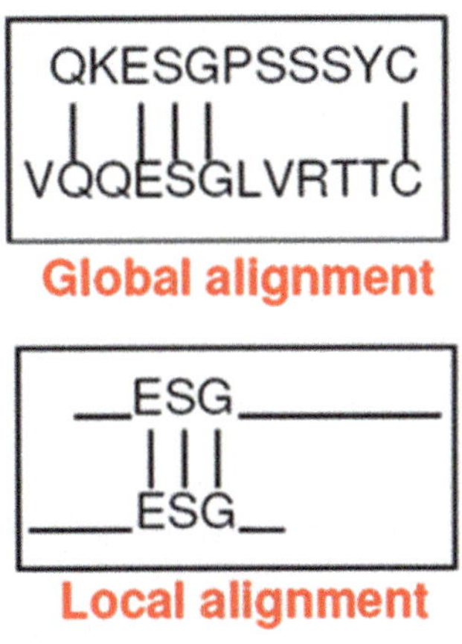

Fig. 4 Schematic representation of global and local alignment

Online Resources to Perform Global and Local Alignment

BLAST (Basic Local Alignment Search Tool) and FASTA are the two most commonly used online tools to find local alignment. BLAST tool is developed by NCBI (National Centre for Biotechnology Information) situated in Bethesda, Maryland, USA [28], whereas FASTA was developed by David J. Lipman and William R. Pearson in 1985. FASTA is the first algorithm which is used to find similar sequences from databases. The algorithm finds optimal local alignments by examining the sequence for small matches referred to as "words." Initially, the scores of segments in which there are multiple word hits are calculated ("init1"). Later the scores of several segments may be summed to generate an "initn" score. An optimized alignment that includes gaps is shown in the output as "opt." The sensitivity and speed of the search are inversely related and controlled by the "k-tup" variable that specifies the size of a "word" [27].

BLAST is a sequence comparison algorithm which is optimized for increased speed to produce best local alignment for a query sequence. The search starts by searching for word of length "W" that scores at least "T" when compared to the query sequence utilizing a substitution matrix. Word hits are then extended in either direction in an attempt to generate an alignment with a score exceeding the threshold of "S." The "T" parameter dictates the speed and sensitivity of the search.

The input required for BLAST is FASTA or Genbank format. Any sequence can be converted to a FASTA format just by adding a ">" symbol in front of it. A variety of BLAST tool are developed, e.g. blastp or protein blast (search protein databases using a protein query), blastn or nucleotide blast (search nucleotide databases using a nucleotide query), blastx (search protein databases using a translated nucleotide query), tblastn (search translated nucleotide databases using a protein query), tblastx (search translated nucleotide databases using a translated nucleotide query). Some other formats for BLAST are also developed like Smart-BLAST, Primer-BLAST, IgBLAST, CDART, MOLE-BLAST, MEGA-BLAST.

The output can be retrieved in various formats, e.g. HTML, plain text, and XML formatting. The default output from NCBI page is HTML. The results are presented in a number of ways like graphical output which shows the hits obtained, a tabular format shows various hits encountered and related data such as E value, percentage query coverage, percentage similarity. The tabular output is the most convenient to understand. Table 1 is showing a list of various online resources for local alignment.

Table 1 List of online tools used for local alignment

Name of site	Web address	References
FASTA program suite	https://fasta.bioch.virginia.edu/fasta_www2/fasta_down.shtml	Pearson and Miller [27]
SIM—Local similarity program for finding alternative alignments	https://web.expasy.org/sim/sim_notes.html	Huang et al. [29] and Huang and Miller [30]
BLAST 2 sequence alignment (BLASTN, BLASTX)	https://blast.ncbi.nlm.nih.gov/Blast.cgi?PAGE=Proteins	Altschul et al. [28]

Substitution Matrices

In Bioinformatics, substitution matrices are defined as the rate at which one amino acid is replaced by another amino acid in a protein sequence during the course of evolution. In other words, it gives the probability of one amino acid being replaced by another one during evolution [31].

What Is the Need of Creating Substitution Matrices?

Let us understand this with the help of an example. Suppose we have two sequences

Seq 1	ATGACTGGA
Seq 2	ATATGA

If we follow the scoring pattern of

Match→1 and Mismatch→0, the two possible alignments with alignment scores are shown below

Alignment 1	Alignment 2
ATGACTGGA	ATGACTGGA
ATA _ _TG_ A Score 5	AT _ A_T_ GA Score 6

So, which one is the best alignment?

The best alignment is the one that best represents the matches among different characters and has maximum alignment score. Hence in this case we can say alignment 2 is the best alignment.

All amino acids are different in chemical nature except the one which belong to the same category like all the charged amino acids have same chemical properties. So, during the course of evolution the probability of substitution of one charged amino acid with an uncharged amino acid is far less than the probability of substitution of a charged amino acid with another charged amino acid. If the former situation occurs, then the simple scoring scheme of 1 for a match and 0 for a mismatch is not enough. So, in order to identify what the probability of substitution of one amino acid is by another amino acid substitution matrices were developed.

Types of Substitution Matrices

There are two basic types of substitution matrices:

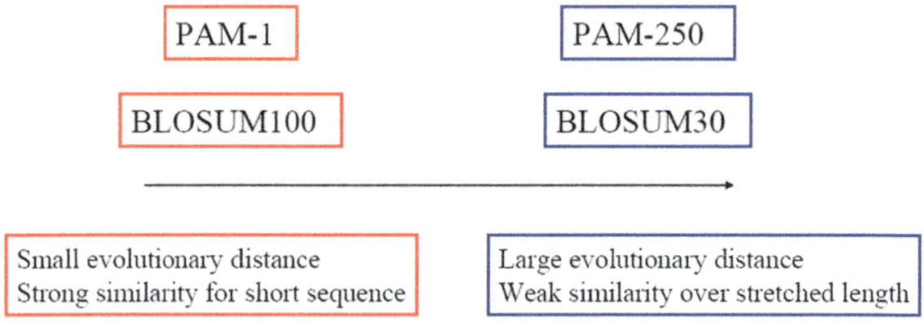

Fig. 5 Relationship between PAM and BLOSSUM matrices

1. Point Accepted Mutation Matrices (PAM): It was developed by Margaret Dayhoff in the 1972s. The basis of PAM construction is to find the probability of mutation of one amino acid by another amino acid during the course of evolution. PAM1 matrix gives us Mutation probability of two amino acid for the evolutionary distance of 1 PAM (i.e., one Accepted Point Mutation per 100 amino acids) [32]. PAM1: A PAM unit is a time period over which 1% of amino acids in a sequence is expected to undergo accepted mutations some of which may occur in the same position. Relation between PAM and BLOSSUM matrix is shown in Fig. 5
2. Blocks Substitution Matrices (BLOSUM): The matrices were created by merging (clustering) all sequences that were more similar than a given percentage into one single sequence and then comparing those sequences (that were all more divergent than the given percentage value) only; thus, reducing the contribution of closely related sequences. The percentage used was appended to the name, giving BLOSUM80, for example, where sequences that were more than 80% identical were clustered. BLOSUM r: the matrix built from blocks with less than r% of similarity—E.g., BLOSUM62 is the matrix built using sequences with less than 62% similarity (sequences with ≥62% identity were clustered) [33].

Differences Between PAM and BLOSUM Matrices

PAM matrices are generally used to score alignments between closely related sequences as compared to BLOSUM which is used to score alignments between evolutionary divergent sequences. PAM matrices are based on global alignment as compared to local alignment on which BLOSUM matrices are based. Figure 5 depicts the relationship between PAM and BLOSUM matrices. Higher PAM matrices like PAM 250 serve the same purpose as BLOSUM 30 which essentially means that these matrices are inversely related. Higher PAM matrices are equivalent to lower BLOSUM matrices [35].

Multiple Sequence Alignment

Due to the advances in modern microbiological and analytical techniques, it is now evident that DNA sequences of various organisms are related and show some level of similarity in their genomes. Some of the widely divergent species have been reported to have conserved sequences of similar genes, often times showing exactly similar functionality, and at some other times they mutate or rearrange themselves to show a completely different function. Hence a lot of genes are present in conserved form in many organisms. A sequence alignment of these genes can reveal the parts which have undergone mutation.

Multiple sequence alignment (MSA) has the potential to reveal the functional and structural similarity of proteins and nucleic acid sequences [34]. The output of MSA reveals homology and the evolutionary relationship between biological sequences. MSA can also be used to identify conserved sequences of protein domains, secondary and tertiary structures of proteins and in some case single amino acid/nucleotides. Computationally, MSA encounters several challenges, first is to find a good alignment for more than two sequences which includes matches, mismatches, and indels and the second is to consider the variation amount in all the sequences for which an alignment needs to be created. A list of few tools used in MSA is shown in Table 2.

A variety of alignment methods comes under the umbrella of MSA to get maximum score without sacrificing on the correctness of alignments.

1. Progressive global alignment: This method was developed by Da-Fei Feng and Doolittle in 1987. This method starts with the two most similar sequences and performing pairwise alignment on them and then it progresses to next one till it reaches the most distantly related. Hence it is known as progressive alignment as it builds an alignment between two most alike sequences and then progressively building the alignment by adding more sequences [36].
2. Iterative method: This method has the same working algorithm as progressive method but after making its first alignment of group of sequences it realigns the results to achieve a more accurate result [35].
3. Alignment which takes in to account the locally conserved patterns found in the same order in the sequences.
4. Statistical and probabilistic methods: These methods can assign probability of occurrence to all feasible combination including matches, mismatches, and indels in order to find out the most appropriate MSA.
5. Consensus method: These methods work to give best MSA from multiple different alignments of same group of sequences, e.g. M-COFFEE [37].

Table 2 Website and program sources for various multiple sequence alignment tools

Name	Source	Reference
Global alignment including progressive CLUSTAL W or CLUSTAL X	https://www.ebi.ac.uk/Tools/msa/clustalo/	Thompson et al. [38] and Higgins et al. [39]
Multiple sequence alignment with hierarchical clustering	http://multalin.toulouse.inra.fr/multalin/multalin.html	
PRALINE	http://ibivu.cs.vu.nl/programs/pralinewww/	Heringa [40]
Iterative and other methods DIALIGN	https://bibiserv.cebitec.uni-bielefeld.de/dialign/	Morgenstern et al. [41]
PRRP	https://www.ncbi.nlm.nih.gov/pmc/articles/PMC145823/	Gotoh [42]
MUSCLE	https://www.ncbi.nlm.nih.gov/pmc/articles/PMC390337/	Edgar [43]
Local alignments of proteins POA Hidden Markov model software	https://academic.oup.com/bioinformatics/article/20/10/1546/237238	Grasso and Lee [44]
MEME website	http://meme-suite.org/	Bailey and Gribskov [45]
MACAW	https://omictools.com/macaw-tool	Schuler et al. [46]
GIBBS	http://www.cs.cmu.edu/~epxing/Class/10810-06/readings/BayesianMotif.pdf	Liu et al. [47]
SAM hidden Markov model	https://compbio.soe.ucsc.edu/sam.html#:~:text=A%20linear%20hidden%20Markov%20model,column%20in%20a%20multiple%20alignment.&text=The%20alignment%20of%20each%20of,creating%20and%20using%20these%20models	Krogh et al. [48]

Phylogenetic Tree

In 1866, Haeckel has coined the term Phylogeny which essentially refers to the evolutionary development of any organism (plant and animal species) or the origin and evolution of a group of plants or animals [49]. The main benefit of phylogenetic tree is it helps in understanding the developmental history (origin of a particular species), assist in understanding epidemiology of infectious diseases. Phylogenetic analysis gives a phylogenetic tree which depicts the relationship among a group of sequences or different species in hierarchical fashion [50].

Phylogenetic trees can be divided in to two broad categories: rooted and unrooted tree. A rooted tree refers to the one which is having a root or a common ancestor or from where all the sequences or species originated; on the other hand, it is very difficult to understand what the hierarchical pattern is in an unrooted tree of species or sequences as shown in Fig. 6.

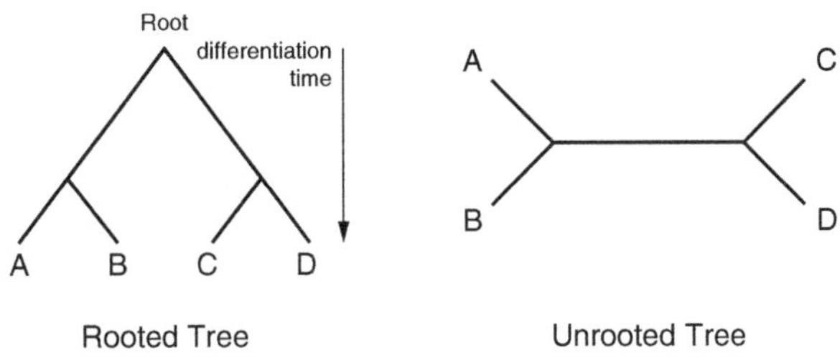

Fig. 6 A rooted tree and an unrooted tree

In a phylogenetic tree every end point is referred to as a node which represents a sequence or specie. So in Figure "a" there are 5 nodes while it is not possible to determine number of nodes as it is not rooted. The point from where a species evolved in to a different one is known as branch point. The branch length in a phylogenetic tree represents how closely one specie is related to its ancestor.

Rooted tree can be of following types:

- Cladogram: Branch length has no meaning.
- Phylogram: Branch length represents evolutionary change.
- Ultra-metric: Branch length represents time and length from root to the leaves are the same.

How to Construct a Phylogenetic Tree?

There are various methods to generate a phylogenetic tree. Few of them are listed below:

a) UPGMA (Unweighted pair group method with arithmetic mean) [50].
b) Neighbor joining [51].
c) Neighbor relation [52].
d) Maximum likelihood approach [52].
e) Transformed distance method [52].

UPGMA Method

Consider 6 sequences for which a phylogenetic tree needs to be constructed.

A: ATCGTGGTACTG
B: CCGGAGAACTAG
C: AACGTGCTACTG
D: ATGGTGAAAGTG
E: CCGGAAAACTTG
F: TGGCCCTGTATC

Step 1: Create a distance matrix simply by checking how much two sequences differ from each other. Like Sequence A and B have 9 differences A and C have 2 difference. The whole matrix is filled following the same process.

	A	B	C	D	E	F
A		9	2	4	9	10
B			9	6	2	10
C				5	9	10
D					6	10
E						10
F						

Step 2: Identify the sequences with fewest difference between them, in this case A–C and B–E are the two pairs which have minimum distance between them.

Step 3: Draw the grouping in the tree; since A–C and B–E have minimum difference between them so they are closely related and hence will be grouped as one.

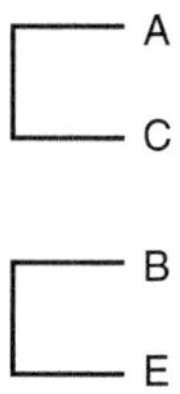

Now, the distance matrix will look like this

	A/C	B	D	E	F
A/C		9	4.5	9	10
B			6	2	10
D				6	10
E					10
F					

The values in the column B and D will be calculated by taking the average of values for A–B, C–B and A–D, C–D, respectively, from the original table.

Step 4: Complete the table with B-E grouped together following the same process.

	A/C	B/E	D	F
A/C		9	4.5	10
B/E			6	10
D				10
F				

Step 5: Repeat steps 2 to 4 until the complete tree is made. The final phylogenetic tree will look like this:

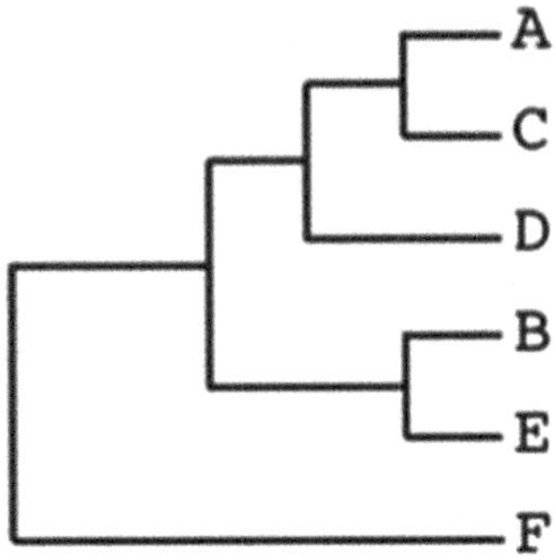

References

1. Huerta M, Downing G, Haseltine F, Seto B, Liu Y. NIH working definition of bioinformatics and computational biology: US National Institute of Health; 2000.
2. Fenstermacher D. Introduction to bioinformatics. J Am Soc Inf Sci Technol. 2005;56(5):440–6.
3. Luscombe NM, Greenbaum D, Gerstein M. What is bioinformatics? A proposed definition and overview of the field. Methods Inf Med. 2001;40(04):346–58.
4. Collins FS, Fink L. The human genome project. Alcohol Health Res World. 1995;19(3):190.
5. Gill SK, Christopher AF, Gupta V, Bansal P. Emerging role of bioinformatics tools and software in evolution of clinical research. Perspect Clin Res. 2016;7(3):115.
6. Mehmood MA, Sehar U, Ahmad N. Use of bioinformatics tools in different spheres of life sciences. J Data Min Genomics Proteomics. 2014;5(2):1.
7. Bettinotti MP, Olsen A, Stroncek D. The use of bioinformatics to identify the genomic structure of the gene that encodes neutrophil antigen NB1, CD177. Clin Immunol. 2002;102(2):138–44.
8. Bell G, Mooers AO. Size and complexity among multicellular organisms. Biol J Linn Soc. 1997;60(3):345–63.
9. Glaessner MF. The first three billion years of life on Earth. J Geogr (Chigaku Zasshi). 1966;75 (6):307–15.
10. Ambrose SH. Paleolithic technology and human evolution. Science. 2001;291(5509):1748–53.
11. Vellai T, Vida G. The origin of eukaryotes: the difference between prokaryotic and eukaryotic cells. Proc R Soc Lond Ser B: Biol Sci. 1999;266(1428):1571–7.
12. Hartman H. The origin of the eukaryotic cell. Specul Sci Technol. 1984;7(2):77–81.
13. Bansal M. DNA structure: revisiting the Watson–Crick. Curr Sci. 2003;85(11):557.
14. Richmond TJ, Davey CA. The structure of DNA in the nucleosome core. Nature. 2003;423 (6936):145–50.
15. Kaplan N, Moore IK, Fondufe-Mittendorf Y, Gossett AJ, Tillo D, Field Y, Segal E, et al. The DNA-encoded nucleosome organization of a eukaryotic genome. Nature. 2009;458 (7236):362–6.
16. Li GW, Xie XS. Central dogma at the single-molecule level in living cells. Nature. 2011;475 (7356):308–15.
17. Moldave K. Eukaryotic protein synthesis. Annu Rev Biochem. 1985;54(1):1109–49.
18. Copley SD, Smith E, Morowitz HJ. A mechanism for the association of amino acids with their codons and the origin of the genetic code. Proc Natl Acad Sci. 2005;102(12):4442–7.

19. Birney E, Clamp M. Biological database design and implementation. Brief Bioinform. 2004;5 (1):31–8.
20. Zou D, Ma L, Yu J, Zhang Z. Biological databases for human research. Genomics Proteomics Bioinformatics. 2015;13(1):55–63.
21. Chang J, Zhu X. Bioinformatics databases: intellectual property protection strategy. J Intellect Prop Rights. 2010;15(6):447–54.
22. Vingron M, Waterman MS. Sequence alignment and penalty choice: review of concepts, case studies and implications. J Mol Biol. 1994;235(1):1–12.
23. Barton GJ. Sequence alignment for molecular replacement. Acta Crystallographica Sect D: Biol Crystallogr. 2008;64(1):25–32.
24. Notredame C. Recent evolutions of multiple sequence alignment algorithms. PLoS Comput Biol. 2007;3(8):e123.
25. Lambert C, Campenhout JM, DeBolle X, Depiereux E. Review of common sequence alignment methods: clues to enhance reliability. Curr Genomics. 2003;4(2):131–46.
26. Likic V. The Needleman-Wunsch algorithm for sequence alignment. Lecture given at the 7th Melbourne Bioinformatics Course, Bi021 Molecular Science and Biotechnology Institute, University of Melbourne; 2008. p. 1–46.
27. Pearson WR, Miller W. Dynamic programming algorithms for biological sequence comparison. In: Methods in enzymology, vol. 210: Academic Press; 1992. p. 575–601.
28. Altschul SF, Gish W, Miller W, Myers EW, Lipman DJ. Basic local alignment search tool. J Mol Biol. 1990;215(3):403–10.
29. Huang X, Hardison RC, Miller W. A space-efficient algorithm for local similarities. Bioinformatics. 1990;6(4):373–81.
30. Huang X, Miller W. A time-efficient, linear-space local similarity algorithm. Adv Appl Math. 1991;12(3):337–57.
31. Altschul SF. Amino acid substitution matrices from an information theoretic perspective. J Mol Biol. 1991;219(3):555–65.
32. Dayhoff MO. A model of evolutionary change in proteins. In: Atlas of protein sequence and structure, vol. 5; 1972. p. 89–99.
33. Henikoff S, Henikoff JG. Amino acid substitution matrices from protein blocks. Proc Natl Acad Sci. 1992;89(22):10915–9.
34. Corpet F. Multiple sequence alignment with hierarchical clustering. Nucleic Acids Res. 1988;16(22):10881–90.
35. Mount DW. Using iterative methods for global multiple sequence alignment. Cold Spring Harb Protoc. 2009;2009(7)., pdb-top44
36. Feng DF, Doolittle RF. Progressive sequence alignment as a prerequisite to correct phylogenetic trees. J Mol Evol. 1987;25(4):351–60.
37. Collingridge PW, Kelly S. MergeAlign: improving multiple sequence alignment performance by dynamic reconstruction of consensus multiple sequence alignments. BMC Bioinformatics. 2012;13(1):117.
38. Thompson JD, Higgins DG, Gibson TJ. CLUSTAL W: improving the sensitivity of progressive multiple sequence alignment through sequence weighting, position-specific gap penalties and weight matrix choice. Nucleic Acids Res. 1994;22(22):4673–80.
39. Higgins DG, Thompson JD, Gibson TJ. Using CLUSTAL for multiple sequence alignments. In: Methods in enzymology, vol. 266: Academic Press; 1996. p. 383–402.
40. Heringa J. Two strategies for sequence comparison: profile-preprocessed and secondary structure-induced multiple alignment. Comput Chem. 1999;23(3–4):341–64.
41. Morgenstern B, Dress A, Werner T. Multiple DNA and protein sequence alignment based on segment-to-segment comparison. Proc Natl Acad Sci. 1996;93(22):12098–103.
42. Gotoh O. Significant improvement in accuracy of multiple protein sequence alignments by iterative refinement as assessed by reference to structural alignments. J Mol Biol. 1996;264 (4):823–38.

43. Edgar RC. MUSCLE: multiple sequence alignment with high accuracy and high throughput. Nucleic Acids Res. 2004;32(5):1792–7.
44. Grasso C, Lee C. Combining partial order alignment and progressive multiple sequence alignment increases alignment speed and scalability to very large alignment problems. Bioinformatics. 2004;20(10):1546–56.
45. Bailey TL, Gribskov M. Methods and statistics for combining motif match scores. J Comput Biol. 1998;5(2):211–21.
46. Schuler GD, Altschul SF, Lipman DJ. A workbench for multiple alignment construction and analysis. Proteins: Struct Funct Bioinf. 1991;9(3):180–90.
47. Liu JS, Neuwald AF, Lawrence CE. Bayesian models for multiple local sequence alignment and Gibbs sampling strategies. J Am Stat Assoc. 1995;90(432):1156–70.
48. Krogh A, Brown M, Mian IS, Sjolander K, Haussler D. Hidden Markov models in computational biology. Applications to protein modeling. J Mol Biol. 1994;235(5):1501–31.
49. Dayrat B. The roots of phylogeny: how did Haeckel build his trees? Syst Biol. 2003;52 (4):515–27.
50. Borriss R, Rueckert C, Blom J, Bezuidt O, Reva O, Klenk HP. Whole genome sequence comparisons in taxonomy. In: Methods in microbiology, vol. 38: Academic Press; 2011. p. 409–36.
51. Saitou N, Nei M. The neighbor-joining method: a new method for reconstructing phylogenetic trees. Mol Biol Evol. 1987;4(4):406–25.
52. Saitou N, Imanishi T. Relative efficiencies of the Fitch-Margoliash, maximum-parsimony, maximum-likelihood, minimum-evolution, and neighbor-joining methods of phylogenetic tree construction in obtaining the correct tree; 1989.

particularly Polymerase Chain Reaction (PCR) based markers and availability of genotyping methods initiated the marker assisted breeding. Marker assisted breeding has speed up the plant breeding and increases the selection intensity and genetic gain. It allowed plant breeders to select trait of interest at early stage of plant which cuts off the required time to release a variety. Marker assisted breeding has been proven highly useful for traits which express at later stage of plant growth, traits which are highly influenced by environment, and other complex traits like male sterility. Further in the genomics era, genomics revolutionized the field of agriculture and crop improvement. It offers a set of tools and techniques to plant breeders which made their job easy and more efficient. The field of genomics and its application is ever increasing in agriculture. Use of genomics information to speed up plant breeding is called as genomics assisted breeding (GAB). Association mapping, Advanced Backcross QTLs (AB-QTLs), Genomic Selection (GS), Multiparent Advance Generation Intercross (MAGIC), Targeting Induced Local Lesions IN Genomes (TILLING), Ecotype-Targeting Induced Local Lesions IN Genomes (Eco-TILING), Marker Assisted Recurrent Selection (MARS), genome-wide diversity studies, etc., are the popular techniques generally used in GAB. These techniques exploit the advances in genomics such as DNA sequencing, gene expression studies, high throughput genotyping and phenotyping, etc., and took plant breeding to its next heights. Numerous genomics data is available in the databases which can be exploited for development of varieties with new and desirable traits such as high yield, biotic and abiotic stress resistance, improved quality and nutrition aspect, and other agronomically important traits.

Today world population is 7.8 billion (https://www.worldometers.info/world-population/) and it is projected to reach 9.6 billion by 2050 [93]. Feeding and nourishing such a growing population need the global crop production to be doubled for next 30 years [78]. Present agriculture is facing problems such as shrinking cultivable land, diminishing irrigation water and its quality, labor unavailability, risk of emerging pest and diseases. All these problems are escalating in the face of global warming and climate change. In such condition there is lot of pressure on crop production for sustainable agriculture. Genetic engineering has the potential to address the issues of crop improvement and food security nevertheless; there is huge rejection of genetically modified (GM) crops by consumers and farmers as well. Sustainable agriculture has the potential to address the issues like food security, climate change, energy requirement of human being, etc. Sustainable agriculture growth can be achieved by use of conventional/past practices of agriculture with the careful use of current genomics technologies. GAB explores the genomics information of crops to accelerate plant breeding and involve in development of high yielding and climate resilient varieties of food crops. Product of GAB is non-GM and has wider acceptance by consumers and farmers. According to FAO, GAB has a great potential to initiate the new "greener revolution" which will feed the ever growing population and at the same time preserve natural resources too [68].

Genomics Advancements

Genome Sequencing

Genome sequencing, the process of determining complete DNA sequence of plant, has been revolutionized in the last two decades and enabled numerous advances in plant biology, crop genomics, molecular systematics, evolutionary genomics, and plant breeding. One example is identification of diagnostic molecular markers for indirect selections for valuable traits during cultivar improvement. Another is unraveling the evolutionary history of important crop species. Till date, the genome information of 1443 plant species are available in the NCBI repository (https://www.ncbi.nlm.nih.gov/genome/browse/#!/eukaryotes/).

The sequencing methods have increased exponentially and the first next-generation sequencing (NGS) became commercially available around 2005. Since then, several sequencing methods have been developed, and these methods are largely grouped into sequencing-by-synthesis (first generation), sequencing-by-ligation (second generation), and single-molecule sequencing (third generation) [30]. The process of sequencing-by-synthesis involves the DNA fragmentation in the appropriate size, ligation to adaptor sequences, and clonal amplification to enhance the fluorescent or chemical signals (Ambardar et al. 2016) [1]. The available platforms for sequencing-by-synthesis are Roche 454 pyrosequencing, Illumina, and Ion Torrent. Sequencing-by-ligation methods use the mismatch sensitivity of DNA ligase to determine the sequence of nucleotides in a given DNA strand [55]. The available platforms for this sequencing are SOLiD and Polonator. Single-molecule sequencing (SMS) also termed produces a detectable signal of nucleotide incorporation via chemiluminescence during DNA sequencing from a single nucleic acid molecule, thus eliminating the need for DNA template amplification, avoids the PCR errors, and biases introduced during template amplification [30]. Currently available platforms for SMS are Pacific Biosciences, PacBio, and nanopore sequencing (Oxford Nanopore Technologies, ONT) [72]. The advent of PacBio and Oxford Nanopore allows the chromosomal level assemblies of plant genomes [7]. The long-read sequencing technologies are often combined with optical mapping and conformation capture, achieving draft genomes of unprecedented contiguity [7, 90]. The sequencing methods with the rapid processing time and high-quality chromosome-scale genome assemblies substantially improve the accuracy of genomic analysis, including gene and regulatory region annotation, genome-wide association studies (GWAS), gene expression quantification, and homologue detection.

Transcriptome Sequencing and Expression Studies

The analysis of gene expression targeting responsive candidate genes or whole transcripts (transcriptome) in specific tissues at a specific time is a functional step in gene characterization. The transcriptome analysis reveals the specific genes involved in the regulation of various physiological responses of plants at any given condition, e.g. under particular stress. In the initial phase, the gene prediction and functional validation have been studied based on the sequencing of cloned single genes and EST sequencing [45, 65] which are good as functional genetic markers, but technically demanding and tedious. Advent of microarray allows the simultaneous analysis of thousands of molecules of unique identity within a single experiment. In microarray, the micro-spots of probe molecules are immobilized in an array format on a solid support and exposed to samples containing the corresponding target molecules [89]. Microarrays have been used to study gene expression and to provide functional analysis of many genes simultaneously [79]. Among the NGS technologies, RNA-Seq is the most powerful tool currently available for comparative transcriptome profiling in plants. The RNA-Seq relies on deep sampling of the transcriptome with many short fragments from a transcriptome and makes computational reconstruction by aligning reads to a reference genome or to each other [59]. Recently, the few studies reported the SMS (PacBio) for transcriptome studies and SMS provides the comprehensive genome annotation including identification of novel genes/isoforms, long non-coding RNAs, and fusion transcripts [110].

Bioinformatics

Bioinformatics applies the power of computers, mathematical algorithms, machine learning, and statistics to solve the biological problems. Bioinformatics helps to generate, analyze, interpret, and store the biological data. The bioinformatics database structured for plant biology may consist of DNA sequences, RNA sequences, protein sequences, molecular markers, phenotypic data, etc. The DNA Data Bank of Japan (DDBJ) [96], GenBank at the National Centre of Biotechnology Information (NCBI) in Bethesda, USA [8, 112], and the European Molecular Biology Laboratory (EMBL) Nucleotide Sequence Database, maintained at the European Bioinformatics Institute (EBI) in the UK [95] are among the largest nucleotide databases. The crop species database repositories are Gramene for cereals, Brassica genome gateway for brassica, MaizeGDB for maize, SoyBase for soybean, Peanutbase for peanut, etc.

Resequencing and SNP Genotyping

As the methodologies leap forward in genome sequencing, it is feasible to sequence multiple cultivars of a crop species. Whole genome resequencing (WGR) is performed on massively parallel sequencing technologies in order to retrieve enough DNA fragments to cover the whole span of the genome of interest in the breeding population/different cultivars of same species. Subsequently WGR links the genome information to economically important traits, enabling the development of DNA markers to assist in breeding. For example, 104 rice cultivars were sequenced by Illumina, identifying millions of polymorphic genomic locations, repeat variations, and single nucleotide polymorphisms (SNPs) and thus providing a rich genomic resource for marker-assisted selection (MAS) [29]. Among the molecular makers, SNPs are most advanced, genome-wide abundant, low-cost, high-throughput, and largely deployed in crop genomics. Various NGS technologies, Illumina GA/Solexa, SOLiDTM, Oxford Nanopore high-throughput sequencing, generated large amount of sequence data in plants; therefore, many new SNPs were identified. Technical advances such as TaqMan and KASP™ (Kompetitive Allele-Specific Polymerase chain reaction) transformed SNP genotyping and brought down the cost of genotyping arrays [4] reduces the genotyping data turnaround time significantly.

Construction of High-Density Genetic Maps

High-density genetic maps are of high importance in the fine mapping for important economic traits and whole genome assembly in plants. NGS and WGR identify a large number of markers, allowing the development of high-density genetic maps construction in plants to map complex features and then identifying genes associated with complex trait/s. Many bioinformatics tools are specially designed to support construction of high-density genetic maps such as $\text{MST}_{\text{MAP,}}$ MSTMap, SEG-Map, etc.

Supremacy of GAB

Conventional breeding mostly relies on selection of individuals based on the phenotype. The individual/lines displaying desired traits are preserved while eliminating the non-rewarding individuals. In the simplest term, phenotype is all about what we concern but that results from the complex process of interaction of genetics and environment. Nonetheless, it is a labor-intensive and time-consuming process that takes multiple generations, even years in case of perennials. Moreover, the specific climate requirements of crops further reduce the efficiency of entire breeding process. With increasing population, breeders aim to keep the demand and crop production neck-to-neck while this is all happening on the same acreage of land. So the future perspective will be the new trends. To this date the genomic selection and precision phenotyping are not really regarded as a new concept but it just keeps gathering new tools to grow about that business. During the 1960s, the era of green revolution, the breeding process was tedious and labor intensive. Now, the DNA sequencing technology has revolutionized the genomic research. For instance, the next-generation sequencing technology involving massively parallel or deep sequencing can sequence an entire genome in a daytime. This enables us to directly look at the genotypes underlying the crop varieties.

The advent of GAB greatly improves the efficiency of breeding process. Like traditional molecular breeding, GAB approach is also employed for identification of novel genes for desired traits in parent plant population, mixing their genotype by traditional breeding approaches and observing the hereditary pattern of desired traits in progeny lines. In its genetics component, one can think about this as the rate of genetic change, means, how much better will be the population in the next year than they are in the current year. In context of breeding cycle this is an integrated process of making new variations and moving through a long and laborious process of evaluation, selection and at last returning the plants for the next cycle. After this, the rate of genetic gain is defined, how long it takes to go through the cycle. With GAB, we aim to improve the genetics and environment. The genetics could be improved by increasing the selection accuracy. This is based upon how close the selection of phenotype is, to the true breeding value. This can be enhanced by taking out experimental error by taking more precise measurements in the right environments. In addition, we may want to reduce the genetic variance over the time. Using the dense genome-wide markers, the genotypic information of a single plant can be extracted in the given phase of plant life-cycle to predict its "total genetic value." This will depend upon the development of superior "training population" which has been precisely genotyped and have quality phenotype to build a genomic selection prediction model that can be applied on the "selection candidate" having only the desired phenotype. This makes GAB inexpensive, time-saving and providing with high-density genotyping. Genotyping by sequencing is getting low per sample. With amazing development in sequencing output, data accuracy also gets improved. With the data get more streamlined, the complications of polyploidy, a typical problem in case of wheat; as well as gene duplication can be tackled that cause ambiguity with hybridization and PCR assays. This can also facilitate polymorphism discovery simultaneously with genotyping.

At its best, GAB enables the selection on single plant early on the breeding process even at seedling stage in unobserved environment. This gives opportunity to make prediction model across different environment at different target location. This also enables us to evaluate relatively large population and even the nursery testing can be made possible before actually sending the breeding lines to the field. This also gives opportunity for maintaining genetic diversity by favorably combining the genetic markers and selecting the lines on preference that maintain rare and low frequency alleles. Apart on a separate note, the plant varieties developed through GAB approaches would face no concern over food safety, government approval, and public acceptance. Unlike GM crops through genetic engineering, they pass on environmental and philosophical relevance.

Technology Involved in GAB

Marker Assisted Selection (MAS)

MAS is the first technique which utilizes molecular tools in breeding. In this method indirect selection for trait of interest usually done based on presence or absence of linked molecular marker. MAS could overcome some of the drawbacks associated with conventional plant breeding and speed up the process. Several markers have been developed and are available to use in many crop species. Restriction Fragment Length Polymorphisms (RFLPs) are the first-generation molecular marker. PCR based markers, viz., Random Amplification of Polymorphic DNAs (RAPDs), Sequence Tagged Sites (STS), Amplified Fragment Length Polymorphisms (AFLPs), Simple Sequence Repeats (SSRs), or microsatellites are included in second-generation markers. Markers based on sequencing like Single Nucleotide Polymorphisms (SNPs) are involved in next-generation molecular markers. Availability of variety of molecular markers, dense genetic maps of different crops and availability of high-throughput genotyping and phenotyping facilities promoted the application of MAS in many crops breeding. Basically three steps are involved in MAS; (1) Identification of marker trait association, (2) Validation of the linked marker, and (3) use of linked marker in selection process. Distance between marker and target gene is the most important factor determining the success of MAS in plant breeding [69]. MAS has been contributing in the field of crop improvement and sustainable agriculture. Readers can refer a very good review written by Collard on MAS [20].

Advanced Backcross QTLs (AB-QTLs)

Despite their inferior phenotype the exotic and wild genotypes are known to possess the genes that can significantly enhance the quality of trait of interest as well as yields. However, these genes and QTLs are more often linked to undesirable traits that reduce the performance of resulting progeny. To overcome such hurdles AB-QTLs method was proposed by Tanksley and Nelson [104], which is an advanced molecular breeding strategy. This technique aims the development of variety by transferring genes/QTLs from unadapted germplasm with the combination of QTL identification. Here, a mapping population is developed by interspecific hybridization with the wild/exotic germplasm and mapping of the desirable trait from the wild genotype is done to transfer in the cultivated background. One of the lacunae of traditional QTL mapping is that detection of QTLs and development of variety are still separate process which requires longer time to develop a variety and further reduces the efficient utilization of information on the QTLs in the process of varietal development. However, AB-QTL helps overcoming this problem by allowing identifying and utilizing QTL simultaneously with the variety

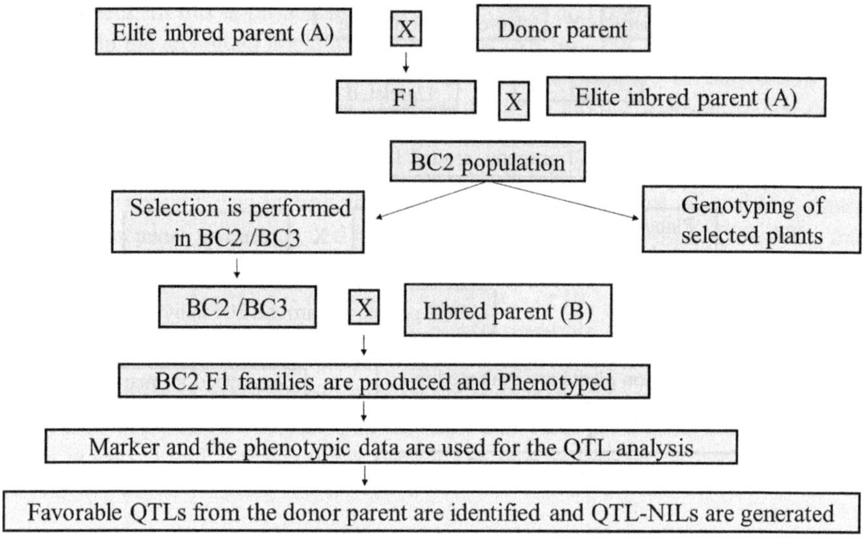

Fig. 2 AB-QTL in cross-pollinated crops

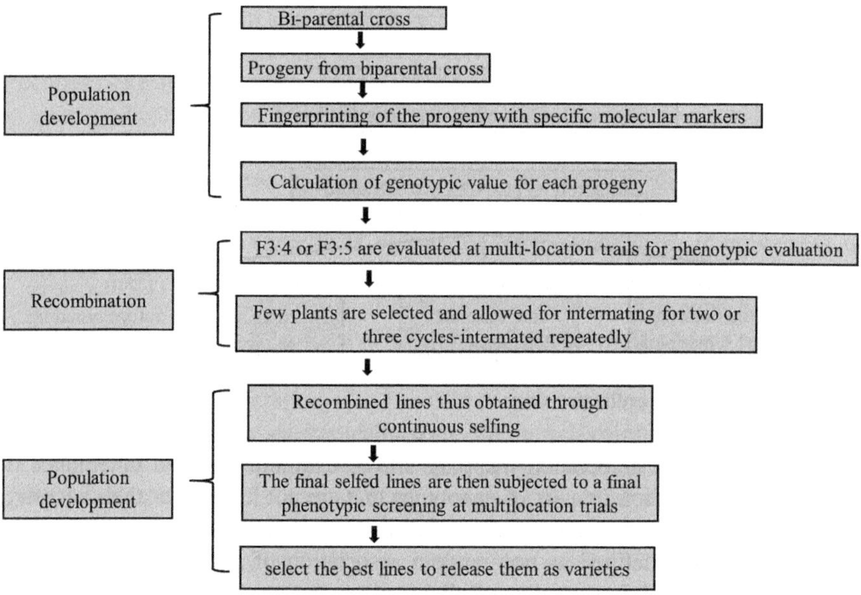

Fig. 3 Simplified schematic representation of MARS

morphologically assessed for the variability of traits of interest by growing them over multilocations and years in replicated trials. Afterwards the samples are subjected to genotyping by using molecular markers. Then the population structure and kinship among the individuals are determined. Linkage disequilibrium between the markers and the trait loci is determined. All these computationally intensive analysis uses different software programs. Thus, linkage disequilibrium analysis can be applied to a number of genomic studies such as MAS. Readers are suggested to refer a very good review written by Myles et al. [64] on association mapping.

Genomic Selection (GS)

The concept of genomic selection was given by Meuwissen et al. [63] to predict complex traits. GS is a variant of marker assisted selection where few molecular markers are used in such a way that all the QTLs in linkage disequilibrium are linked at least with one marker. GS is a form of MAS in which genetic markers covering the whole genome are used so that all QTLs are in linkage disequilibrium with at least one marker. The QTL detection step is skipped in genomic selection method. This technique is efficiently used to design novel breeding methodologies based on molecular genetic markers. It can be very useful in increasing the genetic gain of complex traits per unit time and is cost effective. Availability of high-throughput next-generation sequencing and genotyping facilities serve great purpose in predicting the genomic estimated breeding value accuracies in various crop plants. However, high-throughput phenotyping remains an area of implementation to combine the genetic gains of complex traits by the application of GS. Where MAS is effectively used for the selection of traits governed by fewer genes, genomic selection aims to determine the genetic worth of an individual based on a large set of markers that are distributed throughout the genome. A prediction model for breeding value is developed based on the genotyping and phenotyping of the selected population. The genomic estimated breeding value of all the individuals helps to predict the potential and better performing individuals that can be used as parents' in hybridization program or generation advancement in an ongoing breeding scheme. GS can increase the genetic gain per year compared to phenotypic selection [10, 37, 117]. GS is very useful when phenotypes are difficult or expensive to measure. This allows for (very) early selection, and thus can have economic benefits, as well as faster genetic gain. However, the reference population that is used for genomic selection should be of enough size so that the accurate associations between genotype and phenotype could be estimated. It also needs to be updated by adding new individuals at regular basis because the estimated associations between the SNP and the genes determining the phenotype may be lost due to recombination, mutation, or any other genetic mechanism. GS has been acting as a major support to traditional crop improvement and it is a very important technique to move the GAB into commercial crops with large and complex genomes [70].

Multiparent Advanced Generation Intercross (MAGIC)

This method was first proposed by Mott and coworkers in 2000 in mice. The term was coined by Mackay and Powell in 2008. This is a simple extension of advanced intercross and also described as heterogeneous stock. In the traditional mapping populations, the genomes of the two contrasting parents are combined, which allows the capture of small genomic region affecting the concerned trait. However, in MAGIC population, multiple parents are crossed in different combinations that allow the identification of genes controlling quantitative traits. This is a good method to handle complex pedigree structure and provide opportunity to dissect genomic structure. Multiple inbred founder lines are intermated in a defined pattern for several generations to create resulting inbreeds. This allows formation of population with genomes contributed from multiple parents. The development of MAGIC population requires more initial inputs in terms of careful maintenance of inbreeds and time than the biparental mating. However, the resultant MAGIC population captures much higher degree of diversity; polymorphism and allele frequency [41]. Figure 4 describes a schematic representation of general procedure of development of MAGIC population. A large number of genetic markers across the founders are accumulated through recombination over the populations which are very useful in achieving dense and high-resolution mapping of the genome. Further the use of heterogeneous stocks improves the detection and localization of QTLs compared to an F_2 cross. For detailed information of MAGIC technique readers are directed to go through a review written by Huang et al. [41].

Advantages of MAGIC:

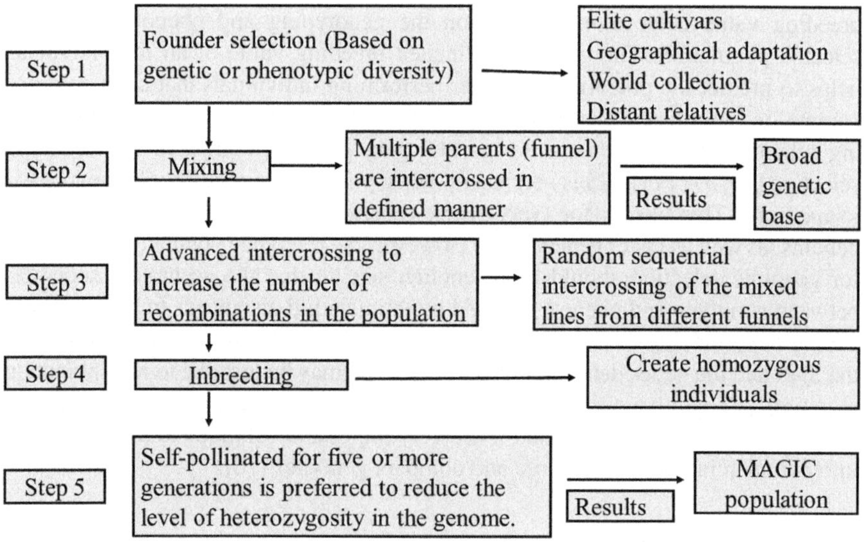

Fig. 4 Steps in development of MAGIC population

1. Large genetic variations are created due to recombinations among different founder lines.
2. Best gene combination can be created for the trait(s) of interest.
3. Superior combinations can be directly used as a variety.
4. Allows identification of allelic variability for complex traits.
5. Provides more precision to identify genetic markers linked to QTLs of traits of interest.

Limitations of MAGIC:

1. Greater investment in terms of cost and time.
2. Limited number of funnels may result relatedness due to shared recombination events and reducing the genetic diversity
3. Requires large-scale phenotyping.
4. Incompatibility between parents may result in relatively a smaller number of progeny developments.
5. Extensive segregation needs to be handled.

Targeting Induced Local Lesions IN Genomes (TILLING) and Ecotype-Targeting Induced Local Lesions IN Genomes (Eco-TILING)

One of the most direct ways to annotate a gene function is to identify an alteration or mutation in the specific gene followed by linking this alteration to a phenotype in the mutant organism or plant. Forward genetic approach is based on the identification of mutant phenotype followed by the gene identification which is responsible for the altered phenotype. However, large population needs to be created and screened for the identification of mutant phenotype which is time and labor consuming. Forward genetic approach in contrast relies on the identification of gene sequence followed by identification of phenotype caused due to gene alteration. This approach has been widely used for gene function annotation. TILLING (Targeting Induced Local Lesions In Genomes) is a general reverse genetic approach that combines chemical mutagenesis with PCR based screening to identify point mutations in the region of interest [54, 62]. TILLING allows analysis of heteroduplex in the individuals in the population that carry point mutation in particular gene (Fig. 5). Tilling is aimed to identify induced single base pair allelic variation or point mutation. While, Eco-tilling aims to identify naturally occurring allelic variations.

Advantages of TILLING:

1. Allows to identify function of thousands of new genes.
2. Allows to identify new mutations or INDELS in genes.
3. Takes advantage of high-throughput screening for nucleotide polymorphisms in a targeted sequence.
4. Can be applied to any species regardless of its genome size and ploidy level.

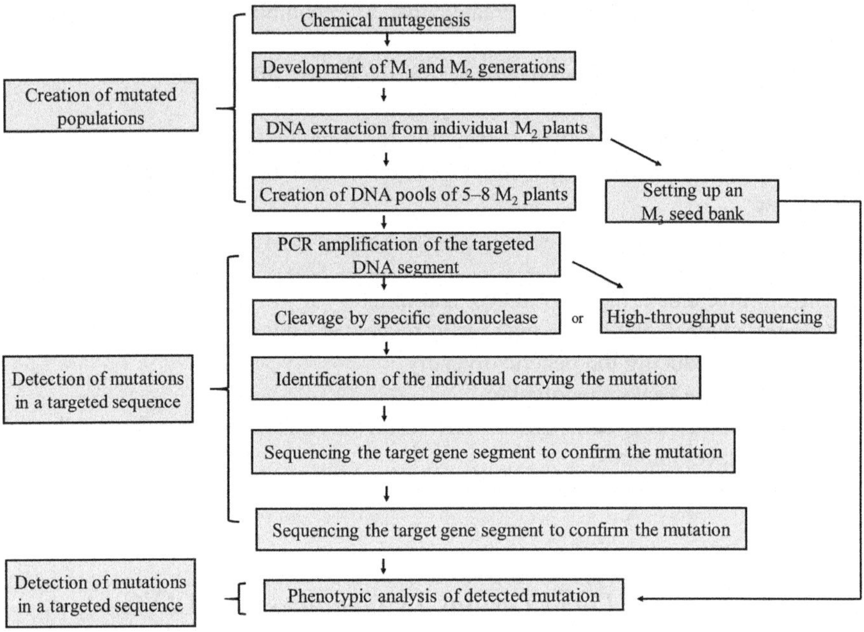

Fig. 5 Simplified schematic representation of TILLING

5. Rapid and low-cost discovery of new alleles that are induced in plants

Successful Applications of GAB

Marker Assisted Selection (MAS)

MAS is widely implemented in several crops for introgressing desired genomic loci for trait improvement [12]. For instance, in rice breeding for improvement of biotic and abiotic stress resistance [19, 35, 46, 67, 91]; wheat breeding [34] for pyramiding genes or QTLs contributing for quality and disease resistance in wheat [32]; Pea breeding [86]; introgression of maize *opaque 2* allele for improving lysine and tryptophan in maize [73] for drought adaptation in maize [81].

AB-QTLs

QTL mapping using advanced backcross populations was thought to be more effective strategy for simultaneous QTL mapping and introgressing the desired genomic loci in to the elite recipient parent. It is being deployed in rice for identification of QTLs for blast resistance [36, 49]; for fine mapping seed dormancy locus in rice [66]; further more applied to wheat breeding programs [39, 40]; in peanut disease resistance improvement [53]; for identification of yield traits in pigeon pea [83]; for identification of kernel width QTL in maize [103].

Association Mapping

The techniques like sequencing of the genomes help in the development of molecular markers [47, 56, 99]. Correspondingly, GWAS aims in identifying the quantitative trait and phenotype association, also genetic diversity among the genotypes which can be useful information for the selection of diverse and potential parents for crop improvement programs. In this context, several reviews are being published in recent times discussing the deployment of GWAS in crop improvement. For instance, Ashokkumar et al. [3] had summarized the genomics enabled breeding approaches in cereal crops like rice, wheat, maize, and pearl millet with an emphasis on biofortification. Similarly, Srivastava et al. [94] had reviewed the prospects of GWAS and GS in pearl millet. Likewise, Battenfield et al. [6] discussed the scope of meta-GWAS in wheat breeding programs. Furthermore, association mapping utilizes existing genetic diversity and thus helps in selecting the effective parental combinations from the natural populations for breeding important traits [42]. Wang et al. [111] have identified novel loci for stripe rust resistance from wheat landrace diversity panel. Puranik et al. [74] have identified nutritional traits in finger millet using 190 accessions of GWAS population.

Genomic Selection

Utilizing whole genome markers proved to be a potential strategy for improving the crop production. It exploits high selection intensity and precision in selecting the elite lines for pre-breeding programs. It has been widely adopted for hybrid rice improvement [21]; rice breeding programs [71]; tomato breeding [14]; maize husk trait improvement [22]; wheat breeding programs [15, 51, 100, 106]; cereal crops improvement [76].

MAGIC

MAGIC are diverse set of permanent mapping populations developed using genomic approaches by International Rice Research Institute (IRRI) for rice breeding programs. The developed lines serve as tailor made diverse germplasm for the breeders to map the quantitative traits and also to exploit them efficiently in rice breeding programs [5]. Descalsota et al. [27] conducted GWAS using MAGIC populations in rice for the biofortification and disease resistance in rice. Likewise MAGIC populations were also developed in wheat, pearl millet, maize, and also in commercial crops like strawberry [26, 82, 84, 107]; Adnan Riaz et al. [80] have identified genomic loci for blotch resistance in wheat using MAGIC populations.

TILLING and Eco-TILLING

Induced mutations in mutagenized populations followed by high-throughput sequencing techniques called TILLING methodology have proved very efficient in identifying the allelic mutations in the target or desired genes of interest which can be harbored for breeding programs. Similarly, Eco-TILLING identifies Single Nucleotide Polymorphisms (SNPs) within a natural population and useful for establishment of marker trait association for breeding interest. This technique has been widely adopted in various crops including cereals. Irshad et al. [43] have critically reviewed the recent advances in TILLING and Eco-TILLING approaches for utilization in breeding programs. In addition, they have summarized the details of target genes studied for allelic variation among different crops using Eco-TILLING. It is suggested that, although TILLING creates random mutations across the genome, it has wide potential when coupled with genome editing technologies. However, this needs to be standardized for several genes of breeding interest and requires improvisation when applied to polyploid genomes like wheat.

Some of the selected examples of the genomics assisted techniques derived products from rice, wheat, and maize have been tabulated in Table 1. All the above discussed techniques have proved their potential in accelerating the breeding programs and also could serve as a valuable resource for dissecting agronomically important traits for crop improvement and sustainability.

Future Prospects

Though the GAB has a great potential for sustainable agriculture, several bottlenecks still impede its direct application. Today with the advent of next-generation sequencing technologies cost of genotyping reduced drastically. Further decrease in genotyping and resequencing cost will accelerate the use of GAB, especially in public plant breeding programs. Genome sequences of many crop plants are available in public domain still, genome of many orphan crops and wild relatives of the crops yet to be sequenced. Availability of genome information of crops and their wild relatives will be the first step toward the use of GAB. High-throughput phenotyping is the prerequisite to accelerate the use of GAB. At present very few phonemics platforms are available in public sector which is a great hindrance for the use of GAB in crop improvement. Initiation of high-throughput phenotyping

Table 1 Successful applications of GAB in crop improvement

Resistance gene/QTL	Marker used	Technique used	Trait improved	References
Rice				
xa13+xa21 (Pusa 1592)	SSR	MAS	Bacterial blight resistance	http://ztmbpd.iari.res.in/technologies/varietieshybrids/cereals/rice/
Pi9 (Pusa Basmati 1637)	SSR	MAS	Blast resistance	Chauhan et al. [16]
xa13+xa21 (Improved Pusa Basmati 1/Pusa 1460)	SSR	MAS	Bacterial blight resistance	Chauhan et al. [16]
xa5 + *xa13* + *Xa21* (Improved Samba Mahsuri)	SSR	MAS	Bacterial blight resistance	Chauhan et al. [16]
Sub1 (Swarna sub1)	SSR	MAS	Submergence	Chauhan et al. [16]
Sub1 (IR64 Sub1)	SSR	MAS	Submergence	Chauhan et al. [16]
Xa4 + *xa5* + *xa13* + *Xa21* (Improved Lalat)	SSR	MAS	Bacterial blight resistance	Chauhan et al. [16]
Xa4 + *xa5* + *xa13* + *Xa21* (Improved Tapaswini)	SSR	MAS	Bacterial blight resistance	Chauhan et al. [16]
Pi2+Pi54 (Pusa Basmati 1609)	SSR	MAS	Blast resistance	https://www.iari.res.in/index.php?option=com_content&view=article&id=649&Itemid=1341
xa13+xa21 (Pusa Basmati 1728)	SSR	MAS	Bacterial blight resistance	https://www.iari.res.in/index.php?option=com_content&view=article&id=649&Itemid=1342
xa13+xa21 (Pusa Basmati 1718)	SSR	MAS	Bacterial blight resistance	https://www.iari.res.in/index.php?option=com_content&view=article&id=649&Itemid=1343
Pi1+Pi54+Pita (Pusa Samba 1850)	SSR	MAS	Blast resistance	https://www.iari.res.in/index.php?option=com_content&

(continued)

Table 1 (continued)

Resistance gene/QTL	Marker used	Technique used	Trait improved	References
				view=article&id=649& Itemid=1344
Pi9, Pi47, Pi48, Pi49, Bph14 and Bph15	SSR	MAS	Blast, BPH	Chen et al. [18]
pi1, pizt, pi5	RFLP	MAS	Blast resistance	Ashkani et al. [2]
Pi1	SSR & ISSR	MAS	Blast resistance	Ashkani et al. [2]
Xa21 & Piz	SSR	MAS	Blast + BLB resistance	Ashkani et al. [2]
PiD1, Pib, Pita, Pi2	SSR	MAS	Blast resistance	Ashkani et al. [2]
PiZ	SSR	MAS	Blast resistance	Ashkani et al. [2]
Xa13, Xa21, Pi54, qSBR11	SSR	MAS	Blast + BLB + sheath blight resistance	Ashkani et al. [2]
Pita	Gene Specific Marker	MAS	Blast resistance	Ashkani et al. [2]
Chrm9 QTL, Xa21, Bph, blast QTL, quality loci	SSR & STS	MABB	SUBMERGENCE + blast + BPH + bacterial leaf blight resistance	Ashkani et al. [2]
Pi1, Pi2, Pi33	SSR	MABB	Blast resistance	Ashkani et al. [2]
Pi1, Pi2, Xa23	SSR	MABB	Blast + BLB resistance	Ashkani et al. [2]
Pizt, Pi54	SSR	MABB	Blast resistance	Ashkani et al. [2]
Pi9	pB8	MABB	Blast resistance	Ashkani et al. [2]
Pi1, Pizt	SSR	MABB	Blast resistance	Ashkani et al. [2]
qSBR11-1, qSBR11-2, qSBR7-1, Pi54	SSR	MABB	BPH + blast resistance	Singh et al. [92]
Bph6 and *Bph12*	SSR	MABB	BPH resistance	Qiu et al. [75]
Bph14, Bph15, Bph18	SSR	MAS	BPH resistance	Hu et al. [38]
Bph14, Bph15, Cry1C, and glufosinate-resistance gene bar.	SSR	MAS	BPH, stem borer, leaf folder resistance, and herbicide resistance	Wan et al. [108]

Bph1 and *Bph2*	STS CAPS MARKERS	MAS	BPH resistance	Sharma et al. [85]
Bph14 and *Bph*15	SSR InDel	MABB	BPH resistance	Wang et al. [109]
Pi2, Pi9, Gm1, Gm4, Sub1, Saltol, xa5, xa13, Xa4 and Xa21 (Improved Lalat)	SSR	MABB	Blast, gall midge resistance, submergence, salinity, BLB resistance	Das and Rao [24]
Six marker trait associations for QTLs identified earlier	SNP markers	GWAS	Starch-related parameters	Biselli [11]
Haplotype analysis of four genes (LOC_Os11g47550, LOC_Os11g47570, LOC_Os11g47590, and LOC_Os11g47610) located in *qAG11* (anaerobic germination)	SNP markers	GWAS	Submergence tolerance	Gao et al. [31]
qDTY12.1, FL478	SSR	MARS	Improvement of IR58025B genetic male sterile for drought and salinity	Suryendra et al. [102]
qRRL2, qSLST1/qRDSW1/qRB1	SNP markers	MAGIC	Salt tolerance	Zhang et al. [115]
GBSSI, SSI, SSIIa, SSIIIa, SBEIa, SBEIIb		Eco-TILLING	Starch synthesis	Irshad et al. [43]
Wheat				
Yr15, Yr62 and *Yr65*	SSR/STS/SCAR/STM/gene specific IC	MAS	Stripe rust resistance	Liu et al. [58]
Yr26, ML91260, Dx5 + Dy10	SSR	MAS	Rust and powdery mildew resistance and quality genes	Zheng et al. [116]
Gpc-B1/Yr36; QGw.ccsu-1A.3; Lr24/Sr24; Lr37/Sr38/Yr17; Yr70/Lr76; Glu-A1–1 and *Glu-A1–2*	Dominant & co dominant markers	MAS	Grain protein content/yellow rust; Grain weight; Leaf rust/stem rust; Leaf rust/stem rust/yellow rust; Yellow rust/leaf rust; Glutenin	Gautam et al. [32]

(continued)

Table 1 (continued)

Resistance gene/QTL	Marker used	Technique used	Trait improved	References
Combined many small effect QTLs	SNP markers	MARS + GWAS	Crown rot resistance	Rahman et al. [77]
63 marker trait associations	SNP markers	GWAS	Root traits	Beyer et al. [9]
23 marker trait associations	SNP markers	GWAS	Agro-morphological traits	Sheoran et al. [87]
Single and multiple marker trait associations	SNP markers	GWAS	Grain micronutrients (Fe, Zn, β-carotene); Grain protein Content and yield traits	Kumar et al. [52]
SNPs for small effect QTLs	SNP markers	GWAS	Fusarium head blight traits	Tessmann et al. [105]
Multiple genetic loci	SNP markers	GWAS	Grain yield and yield components	Sukumaran et al. [97]
23 marker trait associations; pyramiding of minor genes	SNP markers	MARS + GWAS	Crown rot of wheat	Rahman et al. [77]
VRN-A1		Eco-TILLING	Vernalization	Irshad et al. [43]
Pin a, Pin b		Eco-TILLING	Kernel hardness	Irshad et al. [43]
Maize				
qMrdd8	InDel markers	MAS	Rough dwarf disease	Xu et al. [114]
opaque2 allele	SSR	MABB	Improving lysine and tryptophan content	Shetti et al. [88]
opaque 2 (*o2*) gene	SSR	MABC	Improving lysine and tryptophan content	Pukalenthy et al. [73]
SDM QTL	SSR	MABC	Sorghum downy mildew resistance	Sumathi, K et al. [98]
crtRB1 gene (Pusa Vivek QPM 9 Improved)	SSR	MAS	Provitamin-A rich maize hybrid	https://www.iari.res.in/index.php?option=com_content&view=article&id=649&Itemid=1341
(*opaque2* gene) Pusa HM 4 Improved	SSR	MAS	High in lysine (3.62%) and tryptophan (0.91%) as compared to 1.5–2.0% lysine and 0.3–0.4% tryptophan in popular hybrids	https://www.iari.res.in/index.php?option=com_content&view=article&id=649&Itemid=1342

(*opaque2* gene) Pusa HM 8 Improved	SSR	MAS	High in lysine (4.18%) and tryptophan (1.06%) compared to 1.5–2.0% lysine and 0.30.4% tryptophan in popular hybrids	https://www.iari.res.in/index.php?option=com_content&view=article&id=649&Itemid=1343
(*opaque2* gene) Pusa HM 9 Improved	SSR	MAS	High in lysine (2.97%) and tryptophan (0.68%) compared to 1.5–2.0% lysine and 0.30.4% tryptophan in popular hybrids	https://www.iari.res.in/index.php?option=com_content&view=article&id=649&Itemid=1344
(*crtRB1* gene) Pusa Vivek Hybrid 27 Improved	SSR	MAS	High in provitamin-A (5.49 ppm) compared to 0.5–1.5 ppm in popular hybrids under popular storage condition	https://www.iari.res.in/index.php?option=com_content&view=article&id=649&Itemid=1345
(*crtRB1* gene) Pusa HQPM 5 Improved	SSR	MAS	High in provitamin-A (6.77 ppm) compared to 0.5–1.5 ppm in popular hybrids under traditional storage condition	https://www.iari.res.in/index.php?option=com_content&view=article&id=649&Itemid=1346
(*crtRB1* gene) Pusa HQPM 7 Improved	SSR	MAS	High in provitamin-A (7.10 ppm) compared to 0.5–1.5 ppm in popular hybrids under traditional storage condition	https://www.iari.res.in/index.php?option=com_content&view=article&id=649&Itemid=1347
Multiple loci on chromosome 6	SNP markers	MAGIC	Resistance to corn borer	Jiménez-Galindo et al. [50]
A set of 21 QTLs identified	SNP markers	GWAS	Plant height and ear height	Li et al. [57]
qMLN3-108 & *qMLN6-17*	SNP markers	GWAS	Maize chlorotic mottle virus and maize lethal necrosis	Suresh et al. [101]
125 quantitative trait loci (QTLs)	SNP markers	GWAS	Male inflorescence size	Wu et al. [113]
15 quantitative trait loci (QTL)	SNP markers	GWAS	Fusarium ear rot resistance	Chen et al. [17]
Predicted genotypic value from large number of available markers		GS+MARS	Improvement of grain yield and stover-quality traits	Massman et al. [61]

(continued)

Table 1 (continued)

Resistance gene/QTL	Marker used	Technique used	Trait improved	References
Pearl millet				
18 candidate genes	SNP markers	GWAS	Flowering time, plant height, nodal tiller number, and biomass	Diack et al. [28]
QTLs	SNP markers	GWAS	Biomass production in early drought stress conditions and stay-green	Debieu et al. [25]
Hundreds of significant marker trait associations	SNP markers	GWAS	High iron and zinc content in grains	Manwaring et al. [60]
Foxtail millet				
74 marker trait associations	SNP markers	GWAS	Ten nutritional elements	Jaiswal et al. [44]

platforms and phenome projects for main crops will help to achieve full potential of GAB in crop improvement. Indeed phenotyping facility should be affordable. Plant genomics field has been involved in generation of a huge amount of data. There is a great need to manage all the data and make available to the researcher in easy ways with minimum efforts. Bioinformaticians can play crucial role in addressing this problem. Crop specific, trait specific information available in a user friendly way will definitely promote GAB. Further initiatives should be taken to educate and train plant breeders and plant researcher about the use of bioinformatic tools and databases. At present we have enough understanding on inheritance mechanism at genomic level but further knowledge on epigenetic regulation would be helpful to achieve further heights of GAB.

Concluding Remarks

Plant genomics is ever evolving field and adding a new facet or advancement in the field of GAB every day. Under the shadow of global warming and food security GAB is a valuable asset for plant breeders and plant researchers. Due to the inherent limitations of conventional breeding and genetic engineering, GAB is gaining much more importance. GAB has been proving its potential in crop improvement; nevertheless the effort needs to overcome the limitations of GAB which will enhance its widespread application in crop improvement. We believe that extensive use of GAB will help to initiate the new “greener revolution” which will feed the ever growing population and at the same time preserve natural resources of the earth.

References

1. Ambardar S, Gupta R, Trakroo D, Lal R, Vakhlu J, et al. High throughput sequencing: an overview of sequencing chemistry. Indian J Microbiol. 2016;56(4):394–404.
2. Ashkani S, Rafii MY, et al. Molecular breeding strategy and challenges towards improvement of blast disease resistance in rice crop. Front Plant Sci. 2015;6:886.
3. Ashokkumar K, Govindaraj M, et al. Genomics-integrated breeding for carotenoids and folates in staple cereal grains to reduce malnutrition. Front Genet. 2020:11.
4. Ayalew H, Tsang PW, Chu C, Wang J, Liu S, Chen C, Ma XF, et al. Comparison of TaqMan, KASP and rhAmp SNP genotyping platforms in hexaploid wheat. PLoS One. 2019;14(5): e0217222.
5. Bandillo N, Raghavan C, Muyco PA, Sevilla MAL, Lobina IT, Dilla-Ermita CJ, et al. Multi-parent advanced generation inter-cross (MAGIC) populations in rice: progress and potential for genetics research and breeding. Rice. 2013;6(1):11.
6. Battenfield SD, Sheridan JL, Silva LD, et al. Breeding-assisted genomics: applying meta-GWAS for milling and baking quality in CIMMYT wheat breeding program. PLoS One. 2018;13(11):e0204757.
7. Belser C, Istace B, Denis E, Dubarry M, Baurens FC, Falentin C, Deniot G, et al. Chromosome-scale assemblies of plant genomes using nanopore long reads and optical maps. Nat Plants. 2018;4(11):879–87.

8. Benson DA, Karsch-Mizrachi I, Lipman DJ, Ostell J, Sayers EW, et al. GenBank. Nucleic Acids Res. 2009;37(suppl_1):D26–31.
9. Beyer S, Daba S, Tyagi P, et al. Loci and candidate genes controlling root traits in wheat seedlings—a wheat root GWAS. Funct Integr Genomics. 2019;19(1):91–107.
10. Bhat JA, Ali S, Salgotra RK, Mir ZA, DuttaS JV, Tyagi A, Mushtaq M, Jain N, Singh PK, Singh GP, Prabhu KV, et al. Genomic selection in the era of next generation sequencing for complex traits in plant breeding. Front Genet. 2016; https://doi.org/10.3389/fgene.2016.00221.
11. Biselli C, Volante A, Desiderio F, et al. GWAS for starch-related parameters in japonica rice (*Oryza sativa L.*). Plants. 2019;8(8):292.
12. Boopathi NM. Marker-Assisted Selection (MAS). In: Genetic mapping and marker assisted selection. Singapore: Springer; 2020. p. 343–88.
13. Botstein D, White RL, Skolnick M, Davis RW, et al. Construction of a genetic linkage map in man using restriction fragment length polymorphisms. Am J Hum Genet. 1980;32(3):314.
14. Cappetta E, Andolfo G, et al. Accelerating tomato breeding by exploiting genomic selection approaches. Plants. 2020;9(9):1236.
15. Charmet G, Tran LG, et al. BWGS: AR package for genomic selection and its application to a wheat breeding programme. PLoS One. 2020;15(4):e0222733.
16. Chauhan JS, et al. All India coordinated research projects and value for cultivation and use in field crops in India: genesis, outputs and outcomes. Indian J Agric Res. 2016;50:501–10.
17. Chen J, Shrestha R, Ding J, et al. Genome-wide association study and QTL mapping reveal genomic loci associated with Fusarium ear rot resistance in tropical maize germplasm. G3. 2016;6(12):3803–15.
18. Chen Q, Zeng G, et al. Improvement of rice blast and brown planthopper resistance of PTGMS line C815S in two-line hybrid rice through marker-assisted selection. Mol Breed. 2020;40 (2):21.
19. Chukwu SC, Rafii MY, et al. Marker-assisted selection and gene pyramiding for resistance to bacterial leaf blight disease of rice (*Oryza sativa L.*). Biotechnol Biotechnol Equip. 2019;33 (1):440–55.
20. Collard BC, Mackill DJ. Marker-assisted selection: an approach for precision plant breeding in the twenty-first century. Philos Trans R Soc B: Biol Sci. 2008;363(1491):557–72.
21. Cui Y, Li R, et al. Hybrid breeding of rice via genomic selection. Plant Biotechnol J. 2020a;18 (1):57–67.
22. Cui Z, Dong H, et al. Assessment of the potential for genomic selection to improve husk traits in maize. G3. 2020b; https://doi.org/10.1534/g3.120.401600.
23. Dalrymple DG. Development and spread of high-yielding varieties of wheat and rice in the less developed nations (No. 95): Foreign Development Division, Economic Research Service, US Department of Agriculture; 1976.
24. Das G, Rao GJN. Molecular marker assisted gene stacking for biotic and abiotic stress resistance genes in an elite rice cultivar. Front Plant Sci. 2015;6:698.
25. Debieu M, Sine B, Passot S, et al. Response to early drought stress and identification of QTLs controlling biomass production under drought in pearl millet. PLoS One. 2018;13(10): e0201635.
26. Dell'Acqua M, Gatti DM, Pea G, Cattonaro F, Coppens F, Magris G, et al. Genetic properties of the MAGIC maize population: a new platform for high definition QTL mapping in *Zea mays*. Genome Biol. 2015;16(1):1–23.
27. Descalsota GIL, Swamy BP, Zaw H, Inabangan-Asilo MA, Amparado A, et al. Genome-wide association mapping in a rice MAGIC Plus population detects QTLs and genes useful for biofortification. Front Plant Sci. 2018;9:1347.
28. Diack O, Kanfany G, Gueye MC, et al. GWAS unveils features between early-and late-flowering pearl millets. Research Square. 2020. https://doi.org/10.21203/rs.3.rs-25381/v2

29. Duitama J, Silva A, Sanabria Y, Cruz DF, Quintero C, Ballen C, Oard J, et al. Whole genome sequencing of elite rice cultivars as a comprehensive information resource for marker assisted selection. PLoS One. 2015;10(4):e0124617.
30. Egan AN, Schlueter J, Spooner DM, et al. Applications of next-generation sequencing in plant biology. Am J Bot. 2012;99(2):175–85.
31. Gao H, Zhang C, et al. Loci and alleles for submergence responses revealed by GWAS and transcriptional analysis in rice. Mol Breed. 2020;40(8):1–16.
32. Gautam T, Dhillon GS, Saripalli G, et al. Marker-assisted pyramiding of genes/QTL for grain quality and rust resistance in wheat (*Triticum aestivum L.*). Mol Breed. 2020;40:1–14.
33. Gokidi Y, Bhanu AN, Singh MN, et al. Marker assisted recurrent selection: an overview. Adv Life Sci. 2016;5(17):6493–9.
34. Gupta PK, et al. Marker-assisted wheat breeding: present status and future possibilities. Mol Breed. 2010;26(2):145–61.
35. Hasan MM, Rafii MY, et al. Marker-assisted backcrossing: a useful method for rice improvement. Biotechnol Biotechnol Equip. 2015;29(2):237–54.
36. He Y, Jiang H, et al. Identification of blast resistance QTLs based on two advanced backcross populations in rice. Research Square. 2020. https://doi.org/10.21203/rs.2.24273/v2
37. Heffner EL, Lorenz AJ, Jannink JL, Sorrells ME, et al. Plant breeding with genomic selection: gain per unit time and cost. Crop Sci. 2010;50:1681–90.
38. Hu J, Cheng M, et al. Pyramiding and evaluation of three dominant brown planthopper resistance genes in the elite indica rice 93-11 and its hybrids. Pest Manag Sci. 2013;69 (7):802–8.
39. Huang XQ, Cöster H, et al. Advanced backcross QTL analysis for the identification of quantitative trait loci alleles from wild relatives of wheat (*Triticum aestivum L.*). Theor Appl Genet. 2003;106(8):1379–89.
40. Huang XQ, Kempf H, et al. Advanced backcross QTL analysis in progenies derived from a cross between a German elite winter wheat variety and a synthetic wheat (*Triticum aestivum L.*). Theor Appl Genet. 2004;109(5):933–43.
41. Huang BE, Verbyla KL, Verbyla AP, Raghavan C, Singh VK, Gaur P, Leung H, Varshney RK, Cavanagh CR, et al. MAGIC populations in crops: current status and future prospects. Theor Appl Genet. 2015;128(6):999–1017.
42. Ibrahim AK, Zhang L, et al. Principles and approaches of association mapping in plant breeding. Trop Plant Biol. 2020;13:212–24.
43. Irshad A, et al. TILLING in cereal crops for allele expansion and mutation detection by using modern sequencing technologies. Agron. 2020;10(3):405.
44. Jaiswal V, Bandyopadhyay T, et al. Genome-wide association study (GWAS) delineates genomic loci for ten nutritional elements in foxtail millet (*Setaria italica L.*). J Cereal Sci. 2019;85:48–55.
45. Jantasuriyarat C, Gowda M, Haller K, Hatfield J, Lu G, Stahlberg E, Dean RA. Large-scale identification of expressed sequence tags involved in rice and rice blast fungus interaction. Plant Physiol. 2005;138(1):105–15.
46. Jena KK, Mackill DJ. Molecular markers and their use in marker-assisted selection in rice. Crop Sci. 2008;48(4):1266–76.
47. Jia J, Zhao S, et al. *Aegilops tauschii* draft genome sequence reveals a gene repertoire for wheat adaptation. Nature. 2013;496(7443):91–5.
48. Jiang GL, Shi J, Ward RW, et al. QTL analysis of resistance to Fusarium head blight in the novel wheat germplasm CJ 9306. I. Resistance to fungal spread. Theor Appl Genet. 2007;116:3–13.
49. Jiang H, Feng Y, et al. Identification of blast resistance QTLs based on two advanced backcross populations in rice. Rice. 2020;13(1):1–12.
50. Jiménez-Galindo JC, Malvar RA, Butrón A, et al. Mapping of resistance to corn borers in a MAGIC population of maize. BMC Plant Biol. 2019;19(1):431.

51. Kehel Z, Sanchez-Garcia M, et al. Predictive characterization for seed morphometric traits for gene bank accessions using genomic selection. Front Ecol Evol. 2020;8:32.
52. Kumar J, Saripalli G, Gahlaut V, Goel N, et al. Genetics of Fe, Zn, β-carotene, GPC and yield traits in bread wheat (*Triticum aestivum L.*) using multi-locus and multi-traits GWAS. Euphytica. 2018;214(11):219.
53. Kumari V, et al. Utilization of advanced backcross population derived from synthetic amphidiploid for dissecting resistance to late leaf spot in peanut (*Arachis hypogaea L.*). Trop Plant Biol. 2020;13(1):50–61.
54. Kurowska M, Daszkowska-Golec A, Gruszka D, Marzec M, Szurman M, Szarejko I, Maluszynski M, et al. TILLING: a shortcut in functional genomics. J Appl Genet. 2011;52 (4):371–90.
55. Landegren U, Kaiser R, Sanders J, Hood L, et al. A ligase-mediated gene detection technique. Science. 1988;241(4869):1077–80.
56. Li JY, et al. The 3,000 rice genomes project: new opportunities and challenges for future rice research. Gigascience. 2014;3(1):2047-217X.
57. Li X, Zhou Z, Ding J, Wu Y, Zhou B, et al. Combined linkage and association mapping reveals QTL and candidate genes for plant and ear height in maize. Front Plant Sci. 2016;7:833.
58. Liu R, Lu J, Zhou M, et al. Developing stripe rust resistant wheat (*Triticum aestivum L.*) lines with gene pyramiding strategy and marker-assisted selection. Genet Resour Crop Evol. 2020;67(2):381–91.
59. Lowe R, Shirley N, Bleackley M, Dolan S, Shafee T, et al. Transcriptomics technologies. PLoS Comput Biol. 2017;13(5):e1005457.
60. Manwaring HR, Hegarty M, et al. Accessing and dissecting genomic regions for high grain iron and zinc content using GWAS in pearl millet. In: SEB Florence; 8 Apr 2018.
61. Massman JM, et al. Genomewide selection versus marker-assisted recurrent selection to improve grain yield and stover-quality traits for cellulosic ethanol in maize. Crop Sci. 2013;53(1):58–66.
62. McCallum CM, Comai L, Greene EA, Henikoff S, et al. Targeting Induced Local Lesions IN Genomes (TILLING) for plant functional genomics. Plant Physiol. 2000;123:439–42.
63. Meuwissen THE, Hayes BJ, Goddard ME, et al. Prediction of total genetic value using genome-wide dense marker maps. Genetics. 2001;157:1819–29.
64. Myles S, Peiffer J, Brown PJ, Ersoz ES, Zhang Z, Costich DE, Buckler ES, et al. Association mapping: critical considerations shift from genotyping to experimental design. Plant Cell. 2009;21(8):2194–202.
65. Newman T, de Bruijn FJ, Green P, Keegstra K, Kende H, McIntosh L, Retzel E. Genes galore: a summary of methods for accessing results from large-scale partial sequencing of anonymous Arabidopsis cDNA clones. Plant Physiol. 1994;106(4):1241–55.
66. Nguyen T, Fu K, et al. Fine mapping of *qSdr9*, a novel locus for seed dormancy (SD) in weedy rice, and development of NILs with a strong SD allele. Mol Breed. 2020;40(8):1–11.
67. Oladosu Y, Rafii MY, et al. Drought resistance in rice from conventional to molecular breeding: a review. Int J Mol Sci. 2019;20(14):3519.
68. Perez-de-Castro A, Vilanova S, Cañizares J, Pascual L, Blanca JM, Diez MJ, Prohens J, Picó B, et al. Application of genomic tools in plant breeding. Curr Genomics. 2012;13 (3):179–95.
69. Perumalsamy S, Bharani M, Sudha M, Nagarajan P, Arul L, Sarawathi R, et al. Functional marker-assisted selection for bacterial leaf blight resistance genes in rice (*Oryza sativa* L.). Plant Breed. 2010;129:400–6.
70. Poland JA, Rife TW. Genotyping-by-sequencing for plant breeding and genetics. Plant Genome. 2012;5(3):92–102.
71. Prakash P, Arbelaez Velez JD, et al. Empowering global rice breeding programs using genomic selection. [W291] PAG; 2020.

72. Pucker B, Schilbert HM. Genomics and transcriptomics advance in plant sciences. In: Molecular approaches in plant biology and environmental challenges. Singapore: Springer; 2019. p. 419–48.
73. Pukalenthy B, Manickam D, et al. Marker aided introgression of opaque 2 (o2) allele improving lysine and tryptophan in maize (*Zea mays L.*). Physiol Mol Biol Plants. 2020:1–6.
74. Puranik S, Sahu PP, et al. Genome-wide association mapping and comparative genomics identifies genomic regions governing grain nutritional traits in finger millet (*Eleusine coracana L. Gaertn.*). Plants People Planet. 2020; https://doi.org/10.1002/ppp3.10120.
75. Qiu Y, Guo J, et al. Development and characterization of japonica rice lines carrying the brown planthopper-resistance genes *BPH12* and *BPH6*. Theor Appl Genet. 2012;124(3):485–94.
76. Rahim MS, Bhandawat A, Rana N, et al. Genomic selection in cereal crops: methods and applications. In: Accelerated plant breeding, vol. 1. Cham: Springer; 2020. p. 51–88.
77. Rahman M, Davies P, Bansal U, Pasam R, Hayden M, Trethowan R. Marker-assisted recurrent selection improves the crown rot resistance of bread wheat. Mol Breed. 2020;40(3):1–14.
78. Ray DK, Mueller ND, West PC, Foley JA, et al. Yield trends are insufficient to double global crop production by 2050. PLoS One. 2013;8(6):e66428.
79. Rensink WA, Buell CR. Microarray expression profiling resources for plant genomics. Trends Plant Sci. 2005;10(12):603–9.
80. Riaz A, KockAppelgren P, Hehir JG, Kang J, Meade F, Cockram J, et al. Genetic analysis using a multi-parent wheat population identifies novel sources of Septoria Tritici Blotch resistance. Genes. 2020;11(8):887.
81. Ribaut JM, Ragot M. Marker-assisted selection to improve drought adaptation in maize: the backcross approach, perspectives, limitations, and alternatives. J Exp Bot. 2007;58(2):351–60.
82. Sannemann W, Lisker A, Maurer A, Léon J, Kazman E, Cöster H, et al. Adaptive selection of founder segments and epistatic control of plant height in the MAGIC winter wheat population WM-800. BMC Genomics. 2018;19(1):559.
83. Saxena RK, Kale S, et al. Genotyping-by-sequencing and multilocation evaluation of two interspecific backcross populations identify QTLs for yield-related traits in pigeonpea. Theor Appl Genet. 2020;133(3):737–49.
84. Serba DD, Yadav RS. Genomic tools in pearl millet breeding for drought tolerance: status and prospects. Front Plant Sci. 2016;7:1724.
85. Sharma N, et al. Marker-assisted pyramiding of brown planthopper (Nilaparvata lugens Stål) resistance genes *Bph1* and *Bph2* on rice chromosome 12. Hereditas. 2004;140(1):61–9.
86. Sharma A, Sekhon BS, et al. Marker-assisted selection in pea breeding. In: Accelerated plant breeding, vol. 2. Cham: Springer; 2020. p. 137–54.
87. Sheoran S, Jaiswal S, Kumar D, et al. Uncovering genomic regions associated with 36 agro-morphological traits in Indian spring wheat using GWAS. Front Plant Sci. 2019;10:527.
88. Shetti P, et al. Development of lysine and tryptophan rich maize (Zea mays) inbreds employing marker assisted backcross breeding. Plant Gene. 2020:100236.
89. Shi R, Ma W, Wu Q, Zhang B, Song Y, Guo Q, Zheng W. Design and application of 60mer oligonucleotide microarray in SARS coronavirus detection. Chin Sci Bull. 2003;48(12):1165–9.
90. Shi L, Guo Y, Dong C, Huddleston J, Yang H, Han X, Lintner KE. Long-read sequencing and de novo assembly of a Chinese genome. Nat Commun. 2016;7(1):1–10.
91. Singh D, Kumar A, Chauhan P, et al. Marker assisted selection and crop management for salt tolerance: a review. Afr J Biotechnol. 2011;10(66):14694–8.
92. Singh AK, Singh VK, et al. Introgression of multiple disease resistance into a maintainer of Basmati rice CMS line by marker assisted backcross breeding. Euphytica. 2015;203(1):97–107.
93. Skurie J. On World Population Day, Unpacking 9.6 Billion by 2050. National Geographic. National Geographic Society, 11; 2013.
94. Srivastava RK, Singh RB, et al. Genome-wide association studies (GWAS) and genomic selection (GS) in pearl millet: advances and prospects. Front Genet. 2019;10:1389.

95. Sterk P, Kulikova T, Kersey P, Apweiler R, et al. The EMBL nucleotide sequence and genome reviews databases. In: Plant bioinformatics: Humana Press; 2007. p. 1–21.
96. Sugawara H, Ogasawara O, Okubo K, Gojobori T, Tateno Y, et al. DDBJ with new system and face. Nucleic Acids Res. 2007;36(suppl_1):D22–4.
97. Sukumaran S, Dreisigacker S, et al. Genome-wide association study for grain yield and related traits in an elite spring wheat population grown in temperate irrigated environments. Theor Appl Genet. 2015;128(2):353–63.
98. Sumathi K, Ganesan KN, Aarthi P et al. Introgression of QTLs determining sorghum downy mildew (SDM) resistance into elite maize line UMI 79 through marker-assisted backcross breeding (MABC). Australas Plant Pathol. 2020;49:159–165.
99. Sun C, Dong Z, Zhao L, et al. The Wheat 660K SNP array demonstrates great potential for marker-assisted selection in polyploid wheat. Plant Biotechnol J. 2020a;18(6):1354–60.
100. Sun J, Khan M, et al. Genomic selection in wheat breeding. In: Climate change and food security with emphasis on wheat: Academic Press; 2020b. p. 321–30.
101. Suresh LM, Beyene Y, Olsen MS, Makumbi D, et al. Genetic architecture of maize chlorotic mottle virus and maize lethal necrosis through GWAS, linkage analysis and genomic prediction in tropical maize germplasm. Theor Appl Genet. 2019;132(8):2381–99.
102. Suryendra PJ, et al. Marker assisted recurrent selection for genetic male sterile population improvement in rice. Electron J Plant Breed. 2020;11(1):149–55.
103. Tang B, Li Y, et al. Fine mapping and candidate gene analysis of qKW7b, a major QTL for kernel width in maize. Mol Breed. 2020;40(7):1–10.
104. Tanksley SD, Nelson JC. Advanced backcross QTL analysis: a method for the simultaneous discovery and transfer of valuable QTLs from unadapted germplasm into elite breeding lines. Theor Appl Genet. 1996;92(2):191–203.
105. Tessmann EW, Dong Y, Van Sanford DA. GWAS for Fusarium head blight traits in a soft red winter wheat mapping panel. Crop Sci. 2019;59(5):1823–37.
106. Verges VL, Van Sanford DA. Genomic selection at preliminary yield trial stage: training population design to predict untested lines. Agron. 2020;10(1):60.
107. Wada T, Oku K, Nagano S, Isobe S, et al. Development and characterization of a strawberry MAGIC population derived from crosses with six strawberry cultivars. Breed Sci. 2017:17009.
108. Wan B, Zha Z, et al. Development of elite rice restorer lines in the genetic background of R022 possessing tolerance to brown planthopper, stem borer, leaf folder and herbicide through marker-assisted breeding. Euphytica. 2014;195(1):129–42.
109. Wang H, et al. Molecular breeding of rice restorer lines and hybrids for brown planthopper (BPH) resistance using the *Bph14* and *Bph15* genes. Rice. 2016;9(1):53.
110. Wang B, Kumar V, Olson A, Ware D, et al. Reviving the transcriptome studies: an insight into the emergence of single-molecule transcriptome sequencing. Front Genet. 2019;10:384.
111. Wang Y, Yu C, et al. Genome-wide association mapping reveals potential novel loci controlling stripe-rust resistance in a Chinese wheat landrace diversity panel from the southern autumn-sown spring wheat zone. Research Square. 2020. https://doi.org/10.21203/rs.3.rs-22210/v1
112. Wheeler DL, Barrett T, Benson DA, Bryant SH, Canese K, Chetvernin V, Feolo M, et al. Database resources of the national center for biotechnology information. Nucleic Acids Res. 2007;36(suppl_1):D13–21.
113. Wu X, Li Y, Shi Y, Song Y, et al. Joint-linkage mapping and GWAS reveal extensive genetic loci that regulate male inflorescence size in maize. Plant Biotechnol J. 2016;14(7):1551–62.
114. Xu Z, Hua J, Wang F, et al. Marker-assisted selection of qMrdd8 to improve maize resistance to rough dwarf disease. Breed Sci. 2020:19110.
115. Zhang Y, Ponce KS, et al. QTL identification for salt tolerance related traits at the seedling stage in indica rice using a multi-parent advanced generation intercross (MAGIC) population. Plant Growth Regul. 2020:1–9.

116. Zheng W, Li S, et al. Molecular marker assisted gene stacking for disease resistance and quality genes in the dwarf mutant of an elite common wheat cultivar *Xiaoyan22*. BMC Genet. 2020;21:1–8.
117. Zhong S, Dekkers JC, Fernando RL, Jannink JL, et al. Factors affecting accuracy from genomic selection in populations derived from multiple inbred lines: a barley case study. Genet. 2009;182:355–64.

Role of Computational Biology in Sustainable Development of Agriculture

Radheshyam Sharma, Ashish Kumar, and R. Shiv Ramakrishnan

Introduction

Informational technology together with life sciences emerges as a new tempo in last decades. The knowledge generated through the genome sequencing of different plants and organism has grown to a plateau allowing us to use them in a feasible way. Advances in technological interventions opened new avenues for researchers with a favorable condition to concurrently structural and functional study billions of genes, metabolites, and proteins. Thus knowledge generated based on the study has emerged as new scientific domains comprising genomics, metabolomics, proteomics, and bioinformatics. These new emerging fields provide an opportunity for bioprospecting of alleles, discovery of new metabolic pathway, function of novel proteins, etc., for the humankind [1]. In recent past researchers have created superior plants, microbes, animals, food and pharmaceutical products and preserved the existing natural resources and biodiversity. Computational biology which comprises various approaches of bioinformatics to develop algorithms or models and interpret the biological data in order to understand biological systems and inter-relationships. Bioinformatics and computational approaches are indispensable tools to unravel genomics and the molecular systems which lie beneath numerous plant functions. Earlier the research on biological work was mainly focused on laboratory/field experiments only. However, the present scientific era research revolutionized the data interpretation and inference through complex algorithms, computational analysis under *in-silico* conditions.

The Agriculture sector plays an important role in global food security, sustainability and economy of worldwide environment. Agriculture yield has increased during the last two to three decades and will persist to increase as agronomy

R. Sharma (✉) · A. Kumar · R. S. Ramakrishnan
College of Agriculture, Jawaharlal Nehru Krishi Vishwavidyalaya, Jabalpur, India
e-mail: radhebiotech88@jnkvv.org

A. K. Upadhyay et al. (eds.), *Bioinformatics for agriculture: High-throughput approaches*, https://doi.org/10.1007/978-981-33-4791-5_3

re-assorting the enhanced breeding, genome editing and extend new biotechnological-engineered strategies. Since during the experimentation on agricultural crops, different types and huge collection of data are emerged and its interpretation is difficult; thus integrated approach of computational biology and bioinformatics acts a big role to interpret the data in a proper manner. Collection, characterization, and storage of existing plant genetic resource and wisely application of bioinformatics help enrich research in science domain to produce biotic and abiotic resistant crops, biofortification, and post-harvest preservation of produce and improve the quality of livestock.

Computational Biology and Agriculture

Database and computational tools are compressively used in all the biological system to know the biology of an organism. These approaches have made it possible to analyze large datasets to get early understanding and screening of desired targets for detailed experimental purpose. Here we have focused the major thrust area of computational biology.

(a) Crop Plants

Crops are generally derived from their ancestors and domesticated as wild into cultivated form through various natural or artificial process evolutions. When the evolutionary changes occurred in the crop plants through various biological processes, most of the genome remains conserved and less information is available to assess. After the application of database and computational approaches, now we are able to extract and track the desired information from the genome of specific plants. Now the genome information of most of the crops are available freely in the public domain and researchers can use or trace the required data from the domain [2]. Genome sequencing work of different crops is under way. Till date there are several species of crop plants, the genome of which has been mapped completely and available freely for public, for example, arabidopsis, pigeon pea, barley, chick Pea, sorghum, rice, groundnut, maize, wheat, rice, soybean, and the forage legumes (Table 1). Several of these genomes are so large that whole genome sequencing is impractical and researchers are making efforts on comparative genome methods to understand the genomic biology of the crop plants. Cross transferability of gene or part of genome at species or generic level is to be used for comparative information analysis. Similarly the genome organization information of a species with respect to another species is used for transferring superior traits from the wild species to other food crops [3]. Several developed country in the globe provides plant genome sequences and annotations for public domain. Phytozome is an online freely available domain providing genome sequences and annotations of several crop species. Gramene (http://www.gramene.org) is also a freely available public information domain for grass species, which provides a variety of information related to grass genomics, including genome sequences, genome annotations of various grasses

Table 1 List of important published plant genome species

Species name	Size (~Mb)*
Arabidopsis thaliana (Mouse ear cress)	115
Brachypodium distachyon	355
Brassica rapa (Chinese cabbage)	284
Cajanus cajan (Pigeon pea)	833
Carica papaya (Papaya)	372
Cucumis sativus (Cucumber)	203
Fragaria vesca (Woodland strawberry)	240
Glycine max (Soybean)	975
Medicago truncatula (Barrel medic)	241
Malus × domestica (Apple)	881.3
Oryza sativa (Japonica Rice)	372
Panicum virgatum (Switchgrass)	1230
Populus trichocarpa (Poplar)	422.9
Ricinus communis (Castor bean)	400
Pinus taeda (Loblolly pine)	22,180
Solanum tuberosum (Potato)	800
Sorghum bicolor (Sorghum)	730
Theobroma cacao (Cacao)	346
Vitis vinifera (Grapevine)	487
Coriandrum sativum (Coriander)	2118.31
Mangifera indica L. (Mango)	439
Solanum lycopersicum (Tomato)	950
Manihot esculenta (Cassava)	772
Zea mays (Maize)	2500
Triticum aestivum (Wheat)	1700
Hordeum vulgare (Barley)	5100

**Mb* Megabases

[4]. Entrez genome project (http://www.ncbi.nlm.nih.gov/sites/entrez), provides information on genome projects of important agriculture crops such as staple foods, medical plants, fruit trees, forest plants and a number of green alga species.

(b) Renewable Energy

Biofuels that are produced through contemporary processes from biomass provide an ample opportunity to fulfill the global demand as an alternative approach for renewable energy. Plant, algal material, or animal waste based biomass are the finest resources for producing energy by bio-converting it into biofuels such as bio-ethanol, biodiesel, biogas, etc., which are used for various purposes such as blending with petrol and diesel to drive the vehicles and fly the planes, biogas used for any heating purpose. A high biomass containing crop species such as maize (corn), waste potato, switch grass, sugarbeet, rapeseed, oil palm, and agriculture residues like wheat and rice straw, corn stover, orchard prunings, etc., is widely used for biofuel production. Similarly, food waste and portion of food and non-food crops that are not edible can be a good source for producing biofuel. Several government agency and NGO's in India are working in this area to develop ecofriendly and cost-effective biofuel production technology.

Bioinformatics is a tool by which we are able to discover sequence variants in biomass-based crop species to increase the biomass production and recalcitrance. Work on omics approach in improving and development of sustainable biofuel production from microalgae has been carried out and whole genome sequences of more than ten microalgae have been generated [5]. Progress in algal genomics has accelerated the areas to identify several metabolic pathways and genes involved and which is the desirable need for development of genetically engineered micro-algal strains with optimum lipid content [6]. Recently, genome of *Eucalyptus grandis* genome has been sequenced which provided the insights on gene functions. In *Jatropha curcas* comprehensive proteomic studies have been carried out which provide functional information on proteins and their coding genes, metabolic pathways, and targeting of a protein molecule in specific cellular compartments for maximization of biofuel production [7]. Therefore, the use of integrated bioinformatics approach with genomics and proteomics in combination with breeding may be a useful approach to maximize the capability of breeding crop species to be being used as biofuel feedstock and consequently keep increasing the use of renewable energy in present era [8].

(c) Insect Resistance

Insects are the major biotic components of the ecosystem which is either useful or harmful for mankind. Due to their economic importance now researchers focused on the study of insect genomics which helps in the discovery of resistance mechanisms and finding the novel target sites [9]. During last decades various food crops have been developed insect resistance by incorporating single of multiple genes through biotechnological approaches. A well-known example of *Bacillus thuringiensis* is a soil-borne gram positive, soil-dwelling bacterial species whose genome has sequences and used to protect the plants against biotic stress [10]. Researchers used its spores and crystalline insecticidal proteins to control insect pests and genes to incorporate into plants to make it resistant against several lepidopteron, dipteron, coleopteron, hymenopteron pests as well as against nematodes. Till date multiple stacked genes have been successfully transferred to develop genetically modified crops such as cotton, maize, pigeon pea, brinjal, soybean, and potatoes. Plants which have the cry genes from *Bacillus thuringiensis* into their genome, produce a specific toxin. When toxin crystals enter into the digestive system of the insect, their alkaline pH of the digestive tracts denatures the insoluble crystals and makes them in soluble form and thus amenable to being cut with proteases present in the insect gut, which release the toxin from the crystal and transferred into the blood stream and lead to death of insect. These crops are known as Bt crops and these crops are effective against insects by developing resistance against them. Thus bioinformatics plays a vital role for transferring a gene in a precise manner into the plant genome. Now this new approach of the plants to resist insect outbreak may decrease the amount of insecticides being used. As a result the productivity, nutritional value, and quality of plant produce will also increase.

(d) Improve Nutritional Quality

Gene–Diet–Disease interaction of nutritional genomics provides a clear-cut gene interaction study of susceptible genes with dietary interventions of an individual. When the modification is done in the plant genome, then the genotype and phenotype also changed simultaneously which lead to suppress the expression of a trait. Genetic improvement in rice is carried out through transferring genes to increase levels of Pro-vitamin A, zinc, iron, and other desirable micronutrients. Thus this genetic improvement is done through integration of bioinformatic and genomic tools and helped to produce such golden rice which has six times more Pro-vitamin A compared to traditional cultivated rice that can fight against vitamin A deficiencies. Several rice varieties have been developed having rich source Pro-vitamin A that have a profound impact in minimizing the occurrences of blindness and anemia caused by Vitamin A deficiency in African and Asian countries [11]. In maize seeds lysine and tryptophan are limiting amino acid which is essential for the human beings. During the last decades, more effort has been devoted in maize breeding to improve the limiting amino acid content and several quality protein maize varieties have developed. In potato new breeding technology was employed to increase carbohydrates, starch, proteins, minerals, and vitamins. Potato is a stable food and more efforts have been given as compared to other potential food crops. Research work is going on to improve nutritional value of potato tubers, by enhancing *Amaranth Albumin*-1 seed protein content, β-carotene level, vitamin-C content, tuber methionine content, triacylglycerol, amylose content, etc. Consumer-acceptable edible bananas have been developed with increased fruit levels of Pro vitamin A and iron [12]. Researchers have also inserted a gene from yeast into the tomato which leads to increase the stay green ability on vine [13]. New advanced molecular tools like TALEN (transcription activator-like effector nucleases) and CRISPR/Cas9 (clustered regularly interspaced palindromic repeats/CRISPR-associated 9) can be used to develop transgene-free products in a more precise, prompt, and effective way [14]. Therefore, integrated and precise way to be helpful to develop several bio-fortified crops and bioinformatic plays a huge role in schematic data interpretation through various models and algorithm.

(e) Crop Management to Grow in Poorer Soils

Soil is the bin of nutrient resources for seed germination to growth and development of a crop and resource rich soil helps us in producing diverse cropping pattern round the year. But recently deterioration in soil quality has seen due to various factors in several countries which directly affect the cropping pattern and productivity. Alkalinity, salinity, water logging, iron, lead and aluminum toxicity, nutrient deficiency are the major thrust area where reclamation work has to be done to rectify the deterioration of soils. Texture and types of soil are an another factor which affects the crop production. Research work has been carried out to develop crop varieties that have a better tolerance for soil alkalinity, salinity, iron, lead and aluminum toxicities. Several salt-tolerant high-yielding rice varieties have been developed in India for various coastal saline, inland saline and alkaline soils of fragile ecosystems.

Two salt-tolerant varieties, namely, CSR-10 and CSR-11 are the early-maturing dwarf high-yielding rice varieties developed by the Central Soil Salinity Research Institute (CSSRI) and very popular as biological amendments for resource-poor farmers of India [15]. Recently these varieties have a good coverage in poorer soil and adding more rice area in the global production base. Research is in progress to develop crop specific varieties capable of tolerating abiotic and biotic stress [16]. Different varieties have been developed for water logging tolerance in crops like rice, soybean, maize and sugarcane. ICAR-Central Rice Research Institute, Cuttack (India) has developed many high-yielding rice varieties for medium/semi-deep water logged areas in India. Christened as "*CR Dhan 501*" a rice variety is released by the Central Varieties Release Committee in the year 2019 in eastern Indian states. Identification and isolation of quantitative trait loci from wild species or indigenous material is an another way to improve the traits in the plants. Sub 1 QTL for salt tolerance was identified by the researchers and transferred in BR11, Swarna, Samba Mahsuri, IR64, CR1009, and Thadokkam 1 (TDK1) in various countries. Marker assistant molecular breeding with the bioinformatics tools plays a vital role for developing these varieties. The data generated from such intensive research are huge and difficult to manage and analyze manually and bioinformatics helps in a greater amount to solve such problems.

(f) Plant Improvement

Crop improvement through plant breeding tools is a conventional approach to enhance crop productivity under variant climatic conditions without intensifying the application of fertilizers and pesticides. Advancement in the field of omics has provided opportunities for accelerating crop improvement programs. Genomics study is useful to know the genetic and molecular insights of all the biological processes occurring in the plants. Based on the understanding of biological processes, researchers are able to extract the biological information for the development of new crop ideotype with improved quality and other desirable traits [17]. Thus the omics becomes a highly valuable tool for any crop improvement program. Gene expression studies allow us to understand interaction and response of plants with the internal and external stimuli. Thus the data obtained from gene expression analysis act as a crucial tool for developing future breeding strategy and decision management systems [18]. Nowadays researchers also used a reverse breeding approach to detect the function of a gene at phenotypic level and now newly developed genomics approaches are used to exploit the gene pool. Now, identification of genes or transcription factors by functional genomics approaches and transformation technology is used to transfer these genes into crops of interest. Development of gene specific molecular markers and their association with different traits have let to identification of QTL which can be further utilized in molecular breeding through marker assisted selection. In rice *Sub1* is the major QTL identified and transferred for submergence tolerance using molecular breeding approach [19].

(g) Agriculturally Important Microorganism

During the last few decades, research on microorganism has speed up the process and plays pivotal role to understand the various biological process. Several mysteries have been resolved through the advancement in the genomics tools. For rapid and authenticated genomic study of microbes, integrated approach through bioinformatics tools are being used to get the whole genetic architecture of the microorganism and pathogen. Through the bioinformatics researchers are able to check the host–pathogen interaction and how microbes affect the host plants and the knowledge arise could be useful to generate pathogen resistant crop. Although several beneficial microbes are also available in the soil and plant which support the plant to withstand in adverse environment and based on the whole meta transcriptome it is easy to detect the sequence variant of an organism [20, 21]. Biomining, a new field of research that required more emphasis in terms of location specific metal mining. Presently works have been done to detect gold, silver, copper, uranium, and platinum metal through meta genomic sequencing of the contaminated soils [22]. A variety of acidiphilic, chemolithotrophic iron- and sulfur-oxidizing microbes has been identified and used to stimulate the biomining in diverse soils. Polymerase chain reaction based 16S rRNA gene from the environmentally identified samples are being used to detect and identify the microbes in diverse soil and the data generated through the PCR amplification is subsequently analyzed using bioinformatical tools which help in understanding the genetic variation in the microorganism.

In 1994, the U.S. Department of Energy (DOE) started the Microbial Genome Project (MGP) to sequence genomes of microbes which is useful for energy production, biomining, environmental clean-up, toxic degradation, industrial processing, bioremediation, bioleaching, and toxic waste reduction. After studying the genetic information of these organisms, scientists started to understand the insights of the genome and their potentiality to survive in extreme environment.

Corynebacterium glutamicum a gram positive bacterium of high industrial interest and used in the industry for large-scale production of amino acid lysine. *Xanthomonas campestris* another species of bacterium is grown commercially for the production of exopolysaccharide xanthan gum that is used as a viscosifying and stabilizing agent in many industries. *Lactococcus lactis* is another nonpathogenic rod-shaped bacterium for dairy industry. This bacterial strain is critical for manufacturing dairy products such as cheese, yogurt, curd, buttermilk, etc., in addition to dairy industry it is also used to prepare pickled vegetables, wine, bread, beers, sausages, and several other fermented food products. The bacterium *Thermotoga maritima* and the archaeon *Archaeoglobus fulgidus* have great potential for practical applications in industry and bioremediation. Thus, by understanding the physiology and genetic makeup of an organism will certainly help in development of valuable products for the industries.

(h) Accelerate Crop Improvement in a Changing Climate

The climate change and population pressure are eye opening challenges for the researchers to produce sufficient food according to demand. Therefore, it is essential to develop a crop variety which is suitable in adverse climatic condition, give better yield, and adapt in new environment. Advancement in the genomic tools provides an opportunity to accelerate the genomics based breeding in crop plants with precise manner. However, correlating the genomic data with agronomic traits is one of the most challenging tasks and required a skilled person and expertise. For getting suitable results efficient accuracy and handling of the data are also important. Integrated approach of bioinformatics with genomics has the potential to facilitate to maintain food security in climate changing situation through the accelerated production of climate smart food crops [23].

Bioinformatics may also help by the way of whole genome sequencing which can provide novel genes and pathways involved in the reduction of the level of carbon dioxide and other greenhouse gases. Genome sequence of an organism may help to stabilize the global climate change impact. Still very few reports are available in this field and more region-specific research should be conducted for carbon dioxide reduction [24].

(i) Improvement of Plant Resistance Against Abiotic Stresses

The effect of abiotic factors on plants growth and development in a specific environment is a major burning issue during recent past. Such type of stress occurs through various abiotic factors such as high temperature and associated with low water availability, heat, cold, drought salinity, alkalinity, and mineral toxicity and deficiency along with the geographic distribution pattern of plants. These abiotic stresses have made a huge impact on world agriculture, and it has been reported that they reduce the average yields by >50% for major food crops [25]. Generally roots and root hairs are the first line of defense against abiotic stress in plants. If soil is rich in carbon and other nutrients, then plants have a good chance to survive under the stressful conditions. Most of the plants sensitive to external and internal environmental changes but adapt quickly and respond according to the environmental variations. When a group of plants are kept in abiotic stress such as cold, drought then they are prompt to respond uniquely. Therefore, the species become population threatened, endangered, and extinct when abiotic stress occurs. Thus the recent strategy of the researchers is that they identify the defense resistance genes, transcription factors, and metabolic pathway to enhance the immunity and defense mechanism [26, 27]. In various crops through genomics and metabolomic approach scientists have identified the novel defense tolerant genes and pathway against various stresses. Work has been done to develop cereal varieties that have greater tolerance for free aluminum, iron, lead toxicities, soil alkalinity and salinity. Recently in rice, wheat, barley, sugarcane, chickpea, and other many food crops, varieties have been developed that will survive in poorer areas to add to the global production base and continuous progress is to be made to several stresses. Various silico tools have been developed to study expression profiling, physiology, and comparative genomics. The KEGG public database has all metabolic pathways such as the pathway for starch formation, carbohydrate production, and sugar accumulation. Several genes that are involved in proline and ABA production are very important for the development of drought resistance varieties.

(j) Waste Clean-Up

Bacteria, fungi, and other microbes are very useful for bioremediation processes in the environment. Identification, exploitation, and genetic improvement in the microbes are a strategy for increasing their inbuilt potential. For example, a bacterium *Deinococcus radiodurans* has the ability to repair small fragments from chromosomes and damaged DNA by isolating damage segments in specific areas. Scientists have inserted genes from other sources into *D. radiodurans* for environmental clean-up. This recombinant bacterial strain was used to break down organic chemicals, heavy metals, and solvents in radioactive waste sites. Bioinformatics tools and in silico study are important for understanding the mechanisms of bio degradative pathways [28].

Bioinformatics Tools Employed in Agriculture

Bioinformatics tools are playing pivotal role in elucidating the information about different genes present in the genome of an organism. Prediction of functions of different genes and factors affecting their expression was only made possible using the computational *in silico* analysis using bioinformatics software's. The gene level information made it possible for scientist to utilize these sequence data for developing different varieties resistant to abiotic and biotic stresses with improved yield, quality, and other attributes contributing towards the varietal improvement.

1. Sequence Database: Gene and Genome

A biological sequence is a primary biological system object at the molecular level like DNA, RNA, and protein. Recent approaches of sequencing technology have revolutionized the gene/genome sequencing not only with high throughput data generation, but also with its much reduced rates. The core matrix of a bioinformatics system lies in DNA sequence data and several tools have been designed for the annotation, maintenance, and interrogation of sequence information. The largest of the DNA sequence repositories is the International Nucleotide Sequence Database Collaboration (INSDC; http://www.insdc.org), made up of joint efforts of the DNA Data Bank of Japan (DDBJ) at the National Institute for Genetics in Mishima, Japan, GenBank at the National Center of Biotechnology Information (NCBI) (NCBI; http://www.ncbi.nlm.nih.gov/genbank/) in Bethesda, MD, USA, and the European Molecular Biology Laboratory (EMBL) Nucleotide Sequence Database at European Bioinformatics Institute (EMBL-EBI; http://www.ebi.ac.uk/ena) in Hinxton, UK. Various kinds of smaller, specialized databases like different genome browsers, databases of model organisms, molecule- or process-specific databases, etc., have been developed in these repositories for more organized and simplified search of information. These are the three equivalent primary databases of generic DNA sequence data: EMBL, GenBank7, and DDBJ and whenever, any sequence is submitted to one database, it is indexed and distributed automatically to the other databases. These databases are big repositories of sequence information and contain numerous nucleotide sequences stored in them. The INSDC has a uniform policy of no cost and open access to its data [29]. Under the policy, the INSDC captures, conserves, offers, and replaces the comprehensive nucleotide sequence and other correlated information day by day. The INSDC has also launched new services which take care of the richness of the domain, including repositories of raw data from the Trace Archives for Sanger method and Sequence Read Archive (SRA) for next-generation platforms [30], assembly data, experimental design particulars, sample information, taxonomic information, functional annotation and project information. As a conventional data set, assembled sequences and information about their annotations are available from DDBJ [31], the EMBL [32], and GenBank from NCBI [33].

It will not always be pre-requisite, or even suggested, to find the whole of one of the main databases. For example, users may restrict a search to a meticulous organism type such as rodents, vertebrates, or prokaryotes. Further, expressed sequence tags (ESTs), which are small fragments of complementary DNA, may be useful to restrict a database search by excluding ESTs or, alternatively, to search a database consisting entirely of ESTs. These repositories are typically the addresses to visit when someone wishes to retrieve or extract records for further analysis to be performed locally. Complete genome sequences for a variety of crops and animals are now available on the web which made it possible to search a database containing the full gene sequence of a single organism. Web addresses of some selected plant genome database are provided in Table 2.

2. Sequence Analysis Tools

Basic Local Alignment Search Tool (BLAST) is most commonly used similarity search engine [34, 35] for performing database search by using either nucleic acids or protein sequences as a query. It can be freely obtained from NCBI (National Centre for Biotechnology Information) to search GenBank Database on the World Wide Web (http://www.ncbi.nlm.nih.gov/BLAST). Based on the type of query sequence whether it is protein or DNA, a specific program is selected for performing BLAST search. Different BLAST options and their descriptions are provided in Table 3.

For performing BLAST, one of the important steps is to select appropriate database against which the query sequence has to be searched for possible matches. One can make species specific BLAST search by selecting the database from a drop down box. Different types of databases are available in the GenBank. Some of the examples of protein and nucleotide sequences are given in Table 4.

A variety of tools/softwares are now available for different purposes like multiple sequence alignment (MSA), phylogenetic analysis, genetic map constructions, quantitative trait loci identification, etc. A comprehensive list of some basic packages for these specified purposes is given in Table 5.

Table 2 List of important plant genome databases and their web addresses

Plant name	Database	Web address
Arabidopsis thaliana	MATDB (MIPS *A. thaliana* database, Munich, Germ.)	http://mips.helmholtz-muenchen.de/plant/athal/
	TAIR (The Arabidopsis Information Resource, previously AtDB, at Stanford, USA)	https://www.arabidopsis.org/
	KAOS (Kazusa Arabidopsis data Opening Site at Kazusa DNA Research Institute, Japan)	http://www.kazusa.or.jp/kaos/
	Arabidopsis Genome Analysis (Cold Spring Harbor laboratories, USA)	http://www.cshl.org
Triticum aestivum	The Grain genes database	http://www.graingenes.org/
	The ECP/GR wheat database, RICP	http://www.ecpgr.cgiar.org/databases/crops/wheat.htm
	The Field food crop (International rice corporation)	http://www.fao.org/AG/AGP/AGPC/doc/field/Wheat/data/htm
	The TIGR Wheat genome Annotation	http://www.tigr.org/tdb/e2k1/tae1/
Oryza sativa	RGP (Rice Genome Research Programme, Japan)	http://rgp.dna.affrc.go.jp/index.html
	Gramene (Comparative mapping resource for grains)	http://www.gramene.org
	INE (Integrated rice genome explorer: IRGSP, Japan)	http://rgp.dna.affrc.go.jp/giot/INE.html
Zea mays	The TIGR Maize genome Database	http://maize.tigr.org/
	The ECP/GR maize database, RICP BIORES	https://www.mrizp.rs/emdb/default-en.htm
Brassica napus	The ECP/GR *Brassica* database, RICP	http://www.ecpgr.cigar.org/databases/crops/brassica.htm
	The European *Brassica* Database	http://www.actahort.org/books/459/459_28.htn
	Natural Research Environment Council	http://www.brassica.info/ssr/SSRinfo.htm
Medicago truncatula	The TIGR Database	http://www.tigr.org/tdb/e2k1/mta1/
	A model for legume research	http://medicago.org/
	The ECP/GR *Medicago* database, RICP	http://www.ecpgr.cgiar.org/databases/Crops/Medicago.htm
	The Medicago database Query (Agricultural Research Organization of Israel)	http://bioinfo.agri.gov.il/cgi-bin/medicago_query.pl
	Medicago truncatula Sequencing Resources	http://medicago.org/genome/
	Centre for *Medicago* genome Research	http://www.noble.org/medicago/index.htm
	The legume information system (NCGR)	http://www.comparative-legumes.org/

(continued)

Table 2 (continued)

Plant name	Database	Web address
Hordeum vulgare	The plants for a Future	http://www.pfaf.org/database/plants.php?Hordeum+vulgare
	MEROPS *Hordeum vulgare*	http://merops.sanger.ac.uk/cgi-bin/speccards?sp=sp000152&type=P
	TENN Vascular Plants	http://ten.bio.utk.edu/vascular/database/vascular-database.asp
Lycopersicon esculentum	The Tomato Genetics Resource Center	http://tgrc.ucdavis.edu
	The Tomato Expression Database	http://ted.bti.cornell.edu/
	The International Solanaceae	http://www.sgn.cornell.edu/
	Genome Project Tomato database	http://slofly.com/tomatodb/

Table 3 BLAST options, types of query, and database sequence

Program	Compare a query sequence	Against database
Blastn	Nucleotide sequence	Nucleotide sequence
Blastp	Amino acid	Protein sequence
Blastx	A nucleotide sequence translated in all reading frames	Protein sequence
Tblastn	Protein sequence	A nucleotide sequence database dynamically translated in all reading frames
Tblastx	The six-frame translations of a nucleotide sequence	Six-frame translations of a nucleotide sequence database

Conclusions

Integrated approaches of biological science are the major ways to resolve the complex problems in a precise manner. Although several science discipline such as statistics, bioinformatics, computer science, mathematics will help in every field of life. Recently, in the field of agriculture genomic approach provides several benefits in terms of gene prediction, genome annotation, and metabolic pathway mining and comparative evolution of the crop species. Seri bioinformatic database has enriched the research and fostered a surge of new ideas in silk production and improvement. Similarly, other several database and public domain, search engines are the useful tools for the improvement of the specific target crop plant and microorganism. To understand the molecular mechanism and various development pathways of a crop species bioinformatics plays an important role. Thus, based on the insight understanding researchers can produce climate-smart crop plants and lead to produce more food in global base for sustainability and security for the nation.

Table 4 Different type of proteins and nucleotide databases available for performing BLAST search

Database	Description
Protein databases	
Nr	Non-redundant GenBank CDS translations + PDB + SwissProt + PIR + PRF, excluding those in env_nr
Refseq	Protein sequence for NCBI reference project
swissprot	Last major release of the SWISS-PROT protein sequence database (no incremental updates)
Pat	Protein from the Patent division of GenBank
Month	All new or revised GenBank CDS translations + PDB + SwissProt + PIR + PRF, released in the last 30 days
Pdb	Sequence derived from the 3-dimensional structure records from the Protein Data Bank
env_nr	Non-redundant CDS translation from env_nr
Nucleotide database	
Nr	All GenBank + EMBL + DDB + PDB sequences (but no EST, STS, GSS or phase 0,1 or 2 HTGS sequences) No longer "non-redundant" due to computational cost
refseq_mrna	mRNA sequences from NCBI Reference Sequence Project
refseq_genomic	Genomic sequences from NCBI Reference Sequence Project
Est	Database of GenBank + EMBL + DDBJ sequence from EST division
est_human	Human subset of est
est_mouse	Mouse subset of est
est_others	Subset of est other than human or mouse
Gss	Genome Survey Sequence includes single-pass genomic data, exon trapped sequences, and Alu PCR sequences
Htgs	Unfinished High Throughput Genomic Sequences: phases 0, 1, and 2. Finished, phase 3 HTG sequences are in nr
Pat	Nucleotides from the Patent division of GenBank
Pdb	Sequence derived from the 3-dimensional structure records from Protein Data Bank. They are NOT the coding sequences for the corresponding proteins found in the same PDB record
month	All new or revised GenBank + EMBL + DDBJ + PDB sequences released in the last 30 days
alu_repeats	Select Alu repeats from REPBASE, suitable for masking Alu repeats from query sequences
dbsts	Database of Sequence tag site Entries from the STS division of GenBAnk + EMBL + DDBJ
Chromosome	Complete genomes and complete chromosomes from the NCBI References Sequences project. It overlaps with refseq_genomic
Wgs	Assemblies of **W**hole **G**enome **S**hotgun sequences
env_nt	Sequences from environmental samples, such as uncultured bacterial samples isolated from soil or marine samples. This does overlap with nucleotide nr

Table 5 List of bioinformatics software with their features

S. No.	Software	Source	Features	Reference
1	**MAFFT v 7**	http://align.bmr.kyushuu.ac.jp/mafft/online/server/	It can align large number (>2000) of unaligned sequences and can perform rough clustering using N-J and UPGMA approach. It can perform MSA by progressive as well as iterative approaches	Katoh and Standley [36]
2	**MUSCLE**	http://www.drive5.com/muscle/ http://www.ebi.ac.uk/Tools/muscle/index/html	MUSCLE is a freeware used for protein and nucleotide sequences alignment. MUSCLE stands for multiple sequence comparison by log-expectation. It works on the basis of iterative alignment methods with a more accurate measurement to assess the relationship between two sequences	Edgar [37]
3	**CLUSTAL W**	http://www.ebi.ac.uk/Tools/clustalw2/index.html	ClustalW2 is an all-purpose purpose multiple sequence alignment program for DNA or proteins sequences. It gives biologically meaningful multiple sequence alignments of divergent sequences. It calculates the best match for the selected sequences and lines them up so that the identities, similarities, and differences can be seen. Evolutionary relationships can be seen via viewing Cladograms or Phylograms in this software	Thompson et al. [38]
4	**MAP**	http://genome.cs.mtu.edu/map.html	It can align both DNA as well as amino acid sequences. It provides the options for input sequences of files in FASTA format and setting the alignment parameters like match, mismatch scores, gap costs, etc.	Huang [39]
5	**DIALIGN**	http://dialing.gobics.de/ http://bibiserv.techfak.unibielefeld.de/dialign/	It is an iterative alignment based tool freely available to download. It can align both DNA and protein sequences. Sequences can be given as input either in FASTA format	Morgenstern et al. [40] and Morgenstern [41]

(continued)

Table 5 (continued)

S. No.	Software	Source	Features	Reference
			or a FASTA file. DIALIGN constructs pair-wise and multiple alignments by comparing entire segments of the sequences. No gap penalty is used in this software. This approach can be used for both global and local alignment, but it is particularly successful in situations where sequences share only local homologies	
6	**DCA**	http://bibiserv.techfak.unibielefeld.de/dca/	Divide-and-Conquer Multiple Sequence Alignment (DCA) is a program of producing fast, high-quality simultaneous multiple sequence alignment of amino acid, RNA or DNA sequences. The program is based on the DCA algorithm which is a heuristic approach to sum of pairs (SP) optimal alignment	Stoye et al. [42] and Stoye [43]
7	**HMMER**	http://hmmer.janelia.org/	HMMER is a freely available software for protein sequence analysis. It is based on profile hidden Markov models which can be used to do sensitive database searching using statistical description of a sequence family's consensus	Finn et al. [44]
8	**MEME**	http://meme.sdsc.edu/meme/website/intro.html	It is based on Expectation Maximization method. It searches for motifs and then queries them against the database	Bailey et al. [45]

References

1. Debnath M, Pandey M, Bisen P. An omic approach to understand the plants abiotic stress. OMICS. 2011;15:739–62.
2. Gurung PD, Upadhyay AK, Bhardwaj PK, Sowdhamini R, Ramakrishnan U. Transcriptome analysis reveals plasticity in gene regulation due to environmental cues in Primula sikkimensis, a high altitude plant species. BMC Genomics. 2019;20:989.
3. Proost S, Van Bel M, Sterck L, Billiau K, Van Parys T, Van de Peer Y, et al. PLAZA: a comparative genomics resource to study gene and genome evolution in plants. Plant Cell. 2009;21:3718–31.

4. Liang C, Jaiswal P, Hebbard C, Avraham S, Buckler ES, Casstevens T, et al. Gramene: a growing plant comparative genomics resource. Nucleic Acids Res. 2008;36:947–53.
5. Liu B, Benning C. Lipid metabolism in microalgae distinguishes itself. Curr Opin Biotechnol. 2012;24:1–10.
6. Misra N, Panda PK, Parida BK. Agrigenomics for microalgal biofuel production: an overview of various bioinformatics resources and recent studies to link OMICS to bioenergy and bio economy. OMICS. 2013;17:537–49.
7. Maghuly F, Marzban G, Razzazi-Fazeli E, Laimer M. Proteome analyses of *Jatropha curcas*. In: Jankowicz-Cieslak J, Tai T, Kumlehn J, Till B, editors. Biotechnologies for plant mutation breeding. Cham: Springer; 2017.
8. Boyle G. Renewable energy. 2nd ed: Oxford University Press; 2004.
9. Cory JS, Hoover K. Plant-mediated effects in insect-pathogen interactions. Trends Ecol Evol. 2006;21:278–86.
10. Elanchezhian R. ICT for agricultural development in changing climate: Narendra Publishing House; 2012. p. 163–79.
11. Paine JA, Shipton CA, Chaggar S, Howells RM, Kennedy MJ, Vernon G, et al. Improving the nutritional value of Golden Rice through increased pro-vitamin A content. Nat Biotechnol. 2005;23:482–7.
12. Paul J-Y, Harding R, Tushemereirwe W, Dale J. Banana 21: from gene discovery to deregulated golden bananas. Front Plant Sci. 2018;9:528. https://doi.org/10.3389/fpls.2018.00558.
13. Fraser PD, Enfissi E, Bramley PM. Genetic engineering of carotenoid formation in tomato fruit and the potential application of systems and synthetic biology approaches. Arch Biochem Biophys. 2009;483:196–204.
14. Hameed A, Shan-e-Ali Zaidi S, Shakir S, Mansoor S. Applications of new breeding technologies for potato improvement. Front Plant Sci. 2018;9:925. https://doi.org/10.3389/fpls.2018.00925.
15. Mishra B, Singh RK, Senadhira D. Advances in breeding salt-tolerant rice varieties. Adv Rice Genet. 2008:5–7. https://doi.org/10.1142/9789812814319_0002.
16. Wang S, Wan C, Wang Y, Chen H, Zhou Z, Fu H, et al. The characteristics of Na< sup>+, K< sup>+ and free proline distribution in several drought-resistant plants of the Alxa Desert, China. J Arid Environ. 2004;56:525–39.
17. Sharma G, Upadhyay AK, Biradar H, Sonia HS. OsNAC-like transcription factor involved in regulating seed-storage protein content at different stages of grain filling in rice under aerobic conditions. J Genet. 2019;98:18.
18. Langridge P, Fleury D. Making the most of 'omics' for crop breeding. Trends Biotechnol. 2011;29:33–40.
19. Ahmed F, Rafii M, Ismail MR, Juraimi AS, Rahim HA, Asfaliza R, Latif1 MA. Waterlogging tolerance of crops: breeding, mechanism of tolerance, molecular approaches, and future prospects. Hindawi Publishing Corporation. BioMed Res Int. 2013;2013, Article ID 963525, 10 pages. https://doi.org/10.1155/2013/963525.
20. Berg G. Plant-microbe interactions promoting plant growth and health: perspectives for controlled use of microorganisms in agriculture. Appl Microbiol Biotechnol. 2009;84:11–8.
21. Schenk PM, Carvalhais LC, Kazan K. Unraveling plant-microbe interactions: can multispecies transcriptomics help? Trends Biotechnol. 2012;30:177–84.
22. Handelsman J. Metagenomics: application of genomics to uncultured microorganisms. Microbiol Mol Biol Rev. 2004;68:669–85.
23. Batley J, Edwards D. The application of genomics and bioinformatics to accelerate crop improvement in a changing climate. Curr Opin Plant Biol. 2016;30:78–81.
24. Sinha S. Role of bioinformatics in climate change studies. J Sci. 2015;1:1–9.
25. Sahu M, Dehury B, Modi MK, Barooah M. Functional genomics and bioinformatics approach to understand regulation of abiotic stress in cereal crops. In: Crop improvement in the era of climate change Chapter: 19: I.K. International Publishing House Pvt. Ltd; 2014. https://doi.org/10.13140/2.1.4066.1127.

26. Kummerfeld SK, Teichmann SA. DBD: a transcription factor prediction database. Nucleic Acids Res. 2006;34:74–81.
27. Pandey SP, Somssich IE. The role of WRKY transcription factors in plant immunity. Plant Physiol. 2009;150:1648–55.
28. Sadraeian M, Molaee Z. Bioinformatics analyses of Deinococcus radiodurans in order to waste clean-up. In: Environmental and computer science. Second International Conference; 2009. p. 254.
29. Karsch-Mizrachi I, Cochrane G, Nakamura Y. The International Nucleotide Sequence Database Collaboration. Nucleic Acids Res. 2012;40:D33–7.
30. Kodama Y, Shumway M, Leinonen L. On behalf of the International Nucleotide Sequence Database Collaboration The Sequence Read Archive: explosive growth of sequencing data. Nucleic Acids Res. 2012b;40:54–6.
31. Kodama Y, Mashima J, Kaminuma E, Gojobori T, Ogasawara O, Takagi T, Okubo K, Nakamura Y. The DNA Data Bank of Japan launches a new resource, the DDBJ Omics Archive of functional genomics experiments. Nucleic Acids Res. 2012a;40:38–42.
32. Amid C, Birney E, Bower L, Cerdeño-Tárraga A, Cheng Y, Cleland I, Faruque N, Gibson R, Goodgame N, Hunter C, et al. Major submissions tool developments at the European Nucleotide Archive. Nucleic Acids Res. 2012;40:D43–7.
33. Benson DA, Karsch-Mizrachi I, Clark K, Lipman DJ, Ostell J, Sayers EW. GenBank. Nucleic Acids Res. 2012;40:48–53.
34. Altschul S, Gish W, Miller W, Myers E, Lipman D. Basic Local Alignment Search Tool. J Mol Biol. 1990;215:403–10.
35. McGinnis S, Madden TL. BLAST: at the core of a powerful and diverse set of sequence analysis tools. Nucleic Acids Res. 2004;32:20–5.
36. Katoh K, Standley DM. MAFFT multiple sequence alignment software version 7: improvements in performance and usability. Mol Biol Evol. 2013;30(4):772–80.
37. Edgar RC. MUSCLE: a multiple sequence alignment method with reduced time and space complexity. BMC Bioinformatics. 2004;5:113.
38. Thompson JD, Higgins DG, Gibson TJ. CLUSTAL W: improving the sensitivity of progressive multiple sequence alignment through sequence weighting, position-specific gap penalties and weight matrix choice. Nucleic Acids Res. 1994;22:4673–80.
39. Huang X. On global sequence alignment. Comput Appl Biosci. 1994;10:227–35.
40. Morgenstern B, Frech K, Dress A, Werner T. DIALIGN: finding local similarities by multiple sequence alignment. Bioinformatics. 1998;14:290–4.
41. Morgenstern B. Dialign 2: improvement of the segment-to-segment approach to multiple sequence alignment. Bioinformatics. 1999;15(3):211–8.
42. Stoye J, Moulton V, Dress AW. DCA: an efficient implementation of the divide and conquer approach to simultaneous multiple sequence alignment. Comput Appl Biosci. 1997;13:625–6.
43. Stoye J. Multiple sequence alignment with the divide-and-conquer method. Gene. 1998;211:45–56.
44. Finn RD, Clements J, Eddy SR. HMMER web server: interactive sequence similarity searching. Nucleic Acids Res. 2011;39:29–37.
45. Bailey TL, Bodén M, Buske FA, Frith M, Grant CE, Clementi L, Ren J, Li WW, Noble WS. MEME SUITE: tools for motif discovery and searching. Nucleic Acids Res. 2009;37:202–8.

Big Data and Its Analytics in Agriculture

Amit Joshi and Vikas Kaushik

Abbreviations

AI	Artificial intelligence
ANN	Artificial neural network
IoT	Internet of things
ML	Machine learning
NGS	Next generation sequencing

Introduction to Big Data

A great deal of omics information is created in an ongoing decade which flooded the web with transcriptomics, genomics, proteomics, and metabolomics information [1]. The accessibility of biomedical huge information gives a chance to create information driven methodologies in farming and human medicinal services research [2]. As of late, science has become an information intensive science due to colossal datasets created by high-throughput sub-molecular biological tests in different domains including the fields of genomics, transcriptomics, proteomics, and metabolomics. In bioinformatics, the list of components at the genome, transcriptome, proteome, and metabolome levels is bit by bit turning out to be complete and notable to researchers [3]. Bioinformatics finds direct application in the yield improvement programs. It assists specialists in associating hereditary make up with commercial characteristics. Accessibility of complete genomes of various

A. Joshi · V. Kaushik (✉)
Domain of Bioinformatics, School of Bioengineering and Biological Sciences, Lovely Professional University, Phagwara, Punjab, India
e-mail: vikas.14664@lpu.co.in

A. K. Upadhyay et al. (eds.), *Bioinformatics for agriculture: High-throughput approaches*, https://doi.org/10.1007/978-981-33-4791-5_4

monetarily significant crops and headway in facilities and experimentation for high-throughput reads open new insights for crop improvement. Various methodologies like plant genome correlations, hereditary mapping systems, developmental investigations, and so on associated with crop advancement programs are these days conceivable through bioinformatics information examination [4].

Big Data basically alludes to a lot of information which is of organized, semi-organized or unstructured nature. The information pool is voluminous to such an extent that it gets hard for an association to oversee and process it utilizing customary databases and programming methods. Accordingly, huge information not just suggests the huge measure of accessible information; however, it additionally alludes to the whole procedure of social event, putting away, and examining that information. Big data have multiple applications (Fig. 1) in World but in agriculture it plays crucial role like pathogen identification, crop variety improvement, and also in maintaining database for monsoon predictions. It has improved farmer's life much, by providing quality seeds for improving crop yield due to availability of high-throughput transcriptomic data for genetic structures. Every phenotypic character depends on genotypic information, so big data plays critical role in evaluation of disease free crops production.

Agriculture sector is vast in terms of employment generation and providing food security to world population. It includes plant breeding, animal husbandry, fisheries, horticulture, and silviculture practices. Therefore enormous data management for genetic information, meteorological data has to be stored and screened for developmental analysis.

1. Seed Selection and Development of new seed traits,
2. Crop disease and pathogen eradication,
3. Irrigation, Weather, and Climate change assessment,
4. Food security and safety.

Transcriptome Data Analysis in Agriculture

Plants, specifically cereal food crops assume a crucial job in survival of people. In contrast to fauna, floral individuals are sessile and consequently, cannot get away from distressing states, that start from the abiotic stresses (extreme dry, saltiness, and submergence), which may prompt the decrease in yield and nutritional components [5]. Rapid transcriptome analysis for 454-pyrosequencing, Next generation sequencing (Ion torrent sequencing, illumine solexa sequencing), microarray data can be easily assessed for differential gene expression. Next-generation innovations create an accurate measure of genomic information. Effective annotation devices are essential to generate this information amenable to practical genomics investigations. The Mercator pipeline naturally appoints utilitarian terms to protein or nucleotide groupings. It utilizes the MapMan "bin" ontology, which is customized for utilitarian annotation of plant "omics" information [6]. RiceNet v2 (http://www.inetbio.org/ricenet) made by associating varied genomics data and demonstrated the use of the

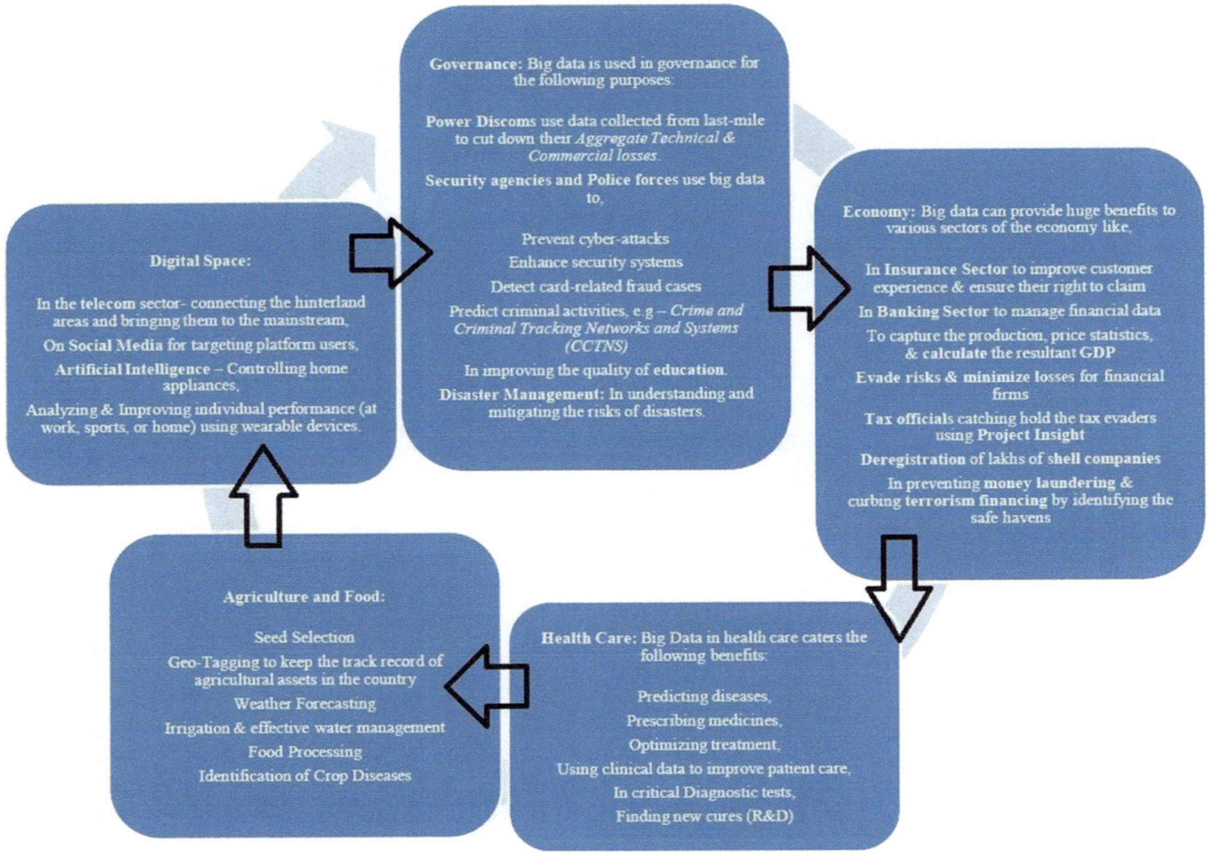

Fig. 1 Applications of Big data in various sectors: Governance, Economy, Health care, Agriculture and food sector, and Digital space

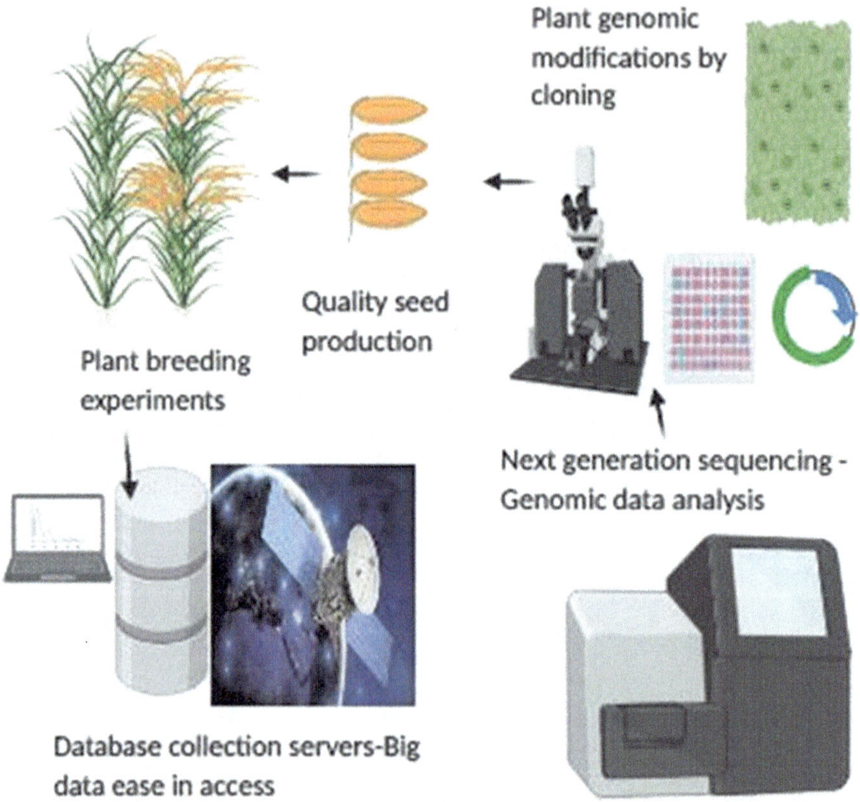

Fig. 2 Big data management: Transcriptomics database generation and flow of information

network in genetic information of rice biotic stress responses and its usefulness for other grass species [7]. Similarly, BBGD454 [8] is a database for blueberry 454 sequencing data and CarrotDB, is a genomic and transcriptomic web-resource for carrot [9]. Pepper EST database for chili pepper transcriptomic data [10].

The USDA Food and Agriculture Cyber informatics and Tools (FACT) activity gives a brilliant chance to supporting high-throughput, computerized information assortment stages, remote detecting, and examination to address open and private agriculture needs. To guarantee that farming datasets are utilized to their most extreme potential, it is imperative to cultivate investigation ability to infer organic experiences for hereditary associations, growth and development expectations, crop quality improvement, and precision agribusiness. In Fig. 2 transcriptomic database creation pipeline is pictorially explained. The sequencing data analyses by RNA-Seq approach assist scientific groups to select gene of interests for agriculture crop. Then, cloning methods help in formulating desirable seeds, and after field investigations desirable crops are screened out. All data of experimentation is uploaded to servers and accessible to Global community for further investigations. The recent approach of RNA sequencing innovation (RNA-Seq) gives a chance to execute exhaustive transcript profiling for heterosis contemplates [11].

Proteomic Data Analysis in Agriculture

Since the accomplishment of sequencing of genomes of living beings, consideration has been engaged to decide the functionality and utilitarian system of proteins by proteome examination. This examination is accomplished by separation and characterization of peptide sequences from proteome, assurance of their capacity and significant system, and development of a suitable database. Numerous upgrades in separation and characterization of proteins, for example, two-dimensional electrophoresis, nano-fluid chromatography, and mass spectrometry, have quickly been accomplished. Some new strategies which incorporate top-down mass spectrometry and tandem affinity purification have risen. These strategies have given the chance of high-throughput investigation of functionality and utilitarian system of proteins in plants. To adapt to the colossal data rising up out of proteome examinations, progressively modern strategies and programming are fundamental [12].

The achievement of cutting edge breading activities relies upon standardized and normalized information management which not just guarantees harmonization of multidimensional information (like genomics, phenotypic, and environmental surrounding) from an association yet in addition encourages network coordination for sharing of data-assets. In present day computerized agriculture, AI, machine learning advancements will keep on producing significant results for farmers and scientific community. In the following period of Novel breeding practices, releasing the intensity of coordinated innovations and big data will empower agricultural production frameworks to characterize genotypes with an optimized phenotype for each unique environmental condition [13]. Intraspecific genetic variation can clarify a high extent of the phenotypes, yet an enormous piece of phenotypic versatility additionally comes from naturally determined transcriptional, post-transcriptional, translational, post-translational, epigenetic and metabolic regulatory processes. Curiously, metabolomics stages are more cost-effective than NGS stages and are definitive for the prediction of dietary benefit or stress control [14].

Databases: Big Data in Agriculture

Big data examination assists in utilizing huge datasets computationally to watch hidden trends, patterns, and result of every strategy. The blend of utilizing savvy cultivating with the big data for various crops can give scrutinized outcomes and the perplexing designs which are not discernible to people in giving the most ideal utilization of agri-resources under the given imperatives [15]. The fundamental necessity to recognize new resistant genetic element to accomplish good quality yield targets is significant for scientific community. Several plant genetic resource servers of public consortiums and private breeding enterprises assist agricultural *in-silico* data for storage, archiving, analyzing, and dissemination plants genetic information. Genetic information storage servers comprise the National-Plant-Germplasm-System (NPGS): a Germplasm Resources Information Network (GRIN) that is an inclusive server storing data on agriculture crops, animal-breeds, microbial organism, and nonvertebrates of commercial importance. Scientific community use such servers can ingress to genomic variedness to produce novel crops that have insects and disease resistance. A European GenBank Integrated System, perpetuated with acquiesce characteristic standards, is also accessible free of cost (http://eurisco.ecpgr.org). The National Genetic Resources Center (NGRC) in Japan executed the NGRC (www.gene.affrc.go.jp/databases_en.php) database for conservation and promotion of agro-biological genomic data. Characterization of proteins and extraction of domains have become an active *in-silico* exploratory area in contemporary era. Different tools are used such as Pfam, PROSITE, MEME, InterProScan, and SAM for proteomic analysis. Expasy server provides majority of protein analysis tools. The sequences of proteins can be retrieved from NCBI Genbank database, structures can be visualized in RCSB-Pdb database. The other primary databases include EMBL, JDB, etc. [16]. In Table 1 all the agriculture big data resource information is enlisted.

Genomics in Agriculture

Numerous agricultural species and their pathogens have sequenced genomes and more are in progress. Farming species give food, fiber, xenotransplant tissues, biopharmaceuticals, and biomedical models. To provide better annotation AgBase (http://www.agbase.msstate.edu) server was designed for structural and functional genomics for agriculture [18]. Plant Reactome (https://plantreactome.gramene.org) is an open-source, relative plant pathway knowledgebase of the Gramene venture. It utilizes Oryza sativa (rice) as a kind of perspective plant categories for manual curation of pathways and stretches out pathway information to another 82 plant species by means of gene-orthology projection utilizing the reactome information model and structure. It presently has 298 reference pathways, including metabolic and transport pathways, transcriptional systems, hormone flagging pathways, and plant formative procedures. Notwithstanding perusing plant pathways, clients can transfer and dissect their omics information, for example, the quality articulation information, and overlay curated or exploratory gene to gene cooperation information to expand pathway information [19]. ZEAMAP (http://www.zeamap.com), a far reaching database fusing various reference genomes, explanations, similar genomics, transcriptomes, open chromatin districts, chromatin collaborations, excellent hereditary variations, phenotypes, metabolomics, hereditary maps, hereditary mapping loci, populace structures and training determination flags among teosinte and maize. ZEAMAP is easy to understand, with the capacity to intelligently

Table 1 Agriculture big data center for plant resource information [17]

Type	Agriculture-data repository	Work information	Web address
Data	BioProject (Biological Project Library)	A public library for collecting descriptive metadata on biological projects	http://bigd.big.ac.cn/bioproject/
	BioSample (Biological Sample Library)	A public library for collecting descriptive metadata on biological materials	http://bigd.big.ac.cn/biosample/
	GSA (Genome Sequence Archive)	A public repository for storing raw sequencing reads generated from different platforms	http://bigd.big.ac.cn/gsa/
Information	GWH (Genome Warehouse)	A centralized repository housing genome-assembly data, including whole genomes, chloroplasts, mitochondria, and plasmids	http://bigd.big.ac.cn/gwh/
	GVM (Genome Variation Map)	A public resource of genome variations, including single-nucleotide polymorphisms and small insertions and deletions	http://bigd.big.ac.cn/gvm/
	GEN (Gene Expression Nebulas)	A database for integrating gene expression profiles	http://bigd.big.ac.cn/gen
	MethBank (Methylation Bank)	A databank of genome-wide DNA methylomes	http://bigd.big.ac.cn/methbank
	NucMap (Nucleosome Positioning Map)	A genome-wide nucleosome positioning map database	http://bigd.big.ac.cn/nucmap/
Knowledge	IC4R (Information Commons for Rice)	A rice knowledge base providing high-quality annotations and integrating multiple omics data	http://ic4r.org/
	PED (Plant Editosome Database)	An expert manually curated knowledge base of plant RNA editosomes	http://bigd.big.ac.cn/ped

incorporate, picture, and cross-reference various distinctive omics datasets [20]. CRISPR/Cas9like genome editing for fruitful results depend on Microarray, and NGS data analysis for crops [21].

Metabolomics in Agriculture

Phenotypic assessment of materials is required on numerous occasions along the crop-rearing pipeline and joining of metabolomics into current practices is supported to significantly abbreviate the improvement time of new assortments, lessen costs, and give unprejudiced phenotypic profiles to approval of hereditary boundaries, and has the capability of being a ground-breaking approach for future accuracy in plants-reproducing. LC-MS, GC-MS used to analyze sequences initially and after that biochemical diversity centered to metabolic activities data generated that stored to databases like (Metabolights, Dataverse, Metabolomics Workbench, Metexplore or Metabolonote) and/or crop specific database such as CassavaBase and MusaBase or PlantCyc [22]. The MetBots framework is a metabolomics accuracy farming platform, for robotized checking of vineyards, giving geo-referenced metabolic pictures that are corresponded and deciphered by an artificial intelligence self-learning framework for supporting precised viticulture. Results can additionally be utilized to investigate the plant metabolic reaction by genome-scale models [23]. A significant test in metabolomics is the absence of dependable explanations for all metabolites distinguished in complex MS as well as NMR information. To address this difficulty, an incorporated UHPLC-QTOF-MS/MS-SPE-NMR framework for higher-throughput metabolite recognizable annotations created, which gives progressed natural setting and upgrades the logical estimation of metabolomics information for understanding agricultural frameworks. This coordinated instrumental technique is computer-server intensive and savvier than regular individual strategies (LC; MS; SPE; NMR). It empowers the concurrent cleansing and distinguishing identification of essential and optional metabolites present in crop varieties [24].

Big Data Analytics

Researchers in the bioscience field have for some time been looking for hereditary variations related with complex phenotypes to propel our comprehension of complex hereditary issues. As a promising instrument for analyzing the hereditary premise of basic maladies, articulation quantitative attribute loci study has pulled in expanding research intrigue. The customary eQTL techniques center around testing the relationship between singular single-nucleotide polymorphisms (SNPs) and quality articulation characteristics.

The Neyman–Pearson (NP) arrangement worldview addresses a significant double characterization issue where clients need to limit type II mistake while controlling sort I blunder under some predetermined level α, generally a modest number. This issue is regularly looked in numerous genomic applications including paired arrangement errands. The phrasing Neyman–Pearson characterization worldview emerges from its association with the Neyman–Pearson worldview in theory testing. The NP worldview is appropriate when one kind of mistake (e.g., type I blunder) is definitely more significant than the other sort (e.g., type II mistake), and clients have a particular objective headed for the previous.

In the period of post-genomics, the errand of improving existing comment is one of the significant tests. Figure 3 describes the significance of genomic information flow to improve crop quality and quantity. The sequenced transcriptome permits to return to the commented on sequenced genome of the comparing life form and improve the current quality models. Furthermore, deceptive explanations proliferate in different databases by relative methodologies of comment, programmed

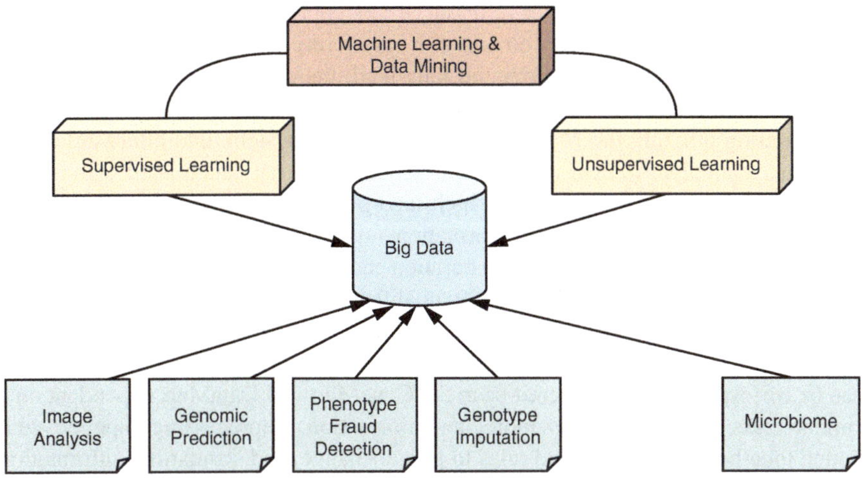

Fig. 3 Big data analytics overview: significance of information flow for crop improvement

comment, and absence of curating power even with enormous information volume. In this interest, re-commented on improved quality models can forestall deceiving basic and utilitarian explanation of qualities and proteins.

Nucleosomes are the structure squares of chromatin and control the physical access of administrative proteins to DNA either straightforwardly or through epigenetic changes. Its situating over the genome leaves a critical effect on the DNA subordinate cycles, especially on quality guideline. In spite of the fact that they structure auxiliary rehashing units of chromatin they contrast from one another by DNA/histone covalent adjustments setting up decent variety in common populaces. Such contrasts incorporate DNA methylation and histone post translational changes happening normally or by the impact of condition. DNA methylation and histone post translational adjustments associate with DNA bringing about quality articulation level changes without modifying the DNA successions and show serious extent of variety among people. Hence, exact planning of nucleosome situating over the genome is fundamental to comprehend the genome guideline. Nucleosome positions and histone borne polymorphism are generally recognized by MNase-Seq and ChIP-CHIP/ChIP-Seq strategies, individually.

In science, the majority of the techniques utilized in genome-wide exploration depend on measurable testing and intended for breaking down a solitary trial dataset. The information blast presented by current genomics innovations expects researcher to reexamine information examination procedures and to make ground-breaking new devices to dissect the information. In late decades, AI has been imagined by life researchers as a superior adaptable learning framework for information driven disclosure. The powerful exhibition of AI has been shown by the Big Data-scale investigation of an aggregation of different information sources from the reference book of DNA components (ENCODE) and model life form reference book of DNA components (modENCODE) ventures in creatures. Hadoop can be based on head of existing elite processing bunches in iPlant, for example, myHadoop, which permits Hadoop occupations to be run on superior registering groups (https://github.com/glennklockwood/myhadoop). Over the long haul, the down to earth steps that Big Data bioinformaticians need to take are: first, overhaul the calculations of existing programming utilizing the MapReduce programming system, and afterward create novel metadata plans to deal with the annotational and trial data related with a plant quality. The trial data specifically, which incorporates information coordination to manufacture a "GeneMart" and information mix to fabricate a "DataMart," may require a half breed of PC and human curation endeavors utilizing an arrangement of bound together marking rules to classify qualities also, datasets dependent on useful depictions and exploratory structures, separately. This strategy is handy to order preparing model sets in an AI examination, in which qualities and information traits can be correspondingly recovered from the GeneMart and DataMart dependent on a similar class names. What is more, an information preprocessing pipeline with bound together calculations and rules to purge, change, and standardize information ought to be created. The heterogeneous idea of natural information ought to be all around thought about when constructing the pipeline on the grounds that, for instance, grouping polymorphic information, quality articulation information, proteomic information, epigenomic information, little RNA information, and record factor restricting information are available in various structures. The focal part of the Big Data investigative stage is a "ToolMart" that provisions an assortment of machine learning models that are arranged dependent on the kinds of learning issues, anticipated results, and types of examined information in plants. These models are pretrained with the models in the GeneMart and DataMart, and are provisioned to clients to use in various parts of AI examinations [25].

Future Scope of Big Data in Agricultural Practices

In modern world meteorological and remote sensing data management [26], genetics database management, pathogenic database construction, and production analytics are integral part of agricultural informatics and there are always improvements required as daily within second's extensive data synthesized at every corner of world. Figure 4 shows the various applications in which data analytics is widely used in agricultural activities. Future predictions will open new dimensions for scientific community to fight with socio-economic issues faced by farmers of developing countries. Main aim of big data repositories is to provide better life, better earning, and better health.

Precision agriculture by deploying IoT, wireless sensor, embedded system, RFID, and satellite networks, make big data analysis possible [27]. Scientific activities centric to omics, *in-silico* and computational chemistry of agricultural crops and animal husbandry and breeding empower the researchers and associations to maintain food security for the global populace and ameliorate the food and fodder quality standards. Advancement in agricultural production can serve as a significant task for

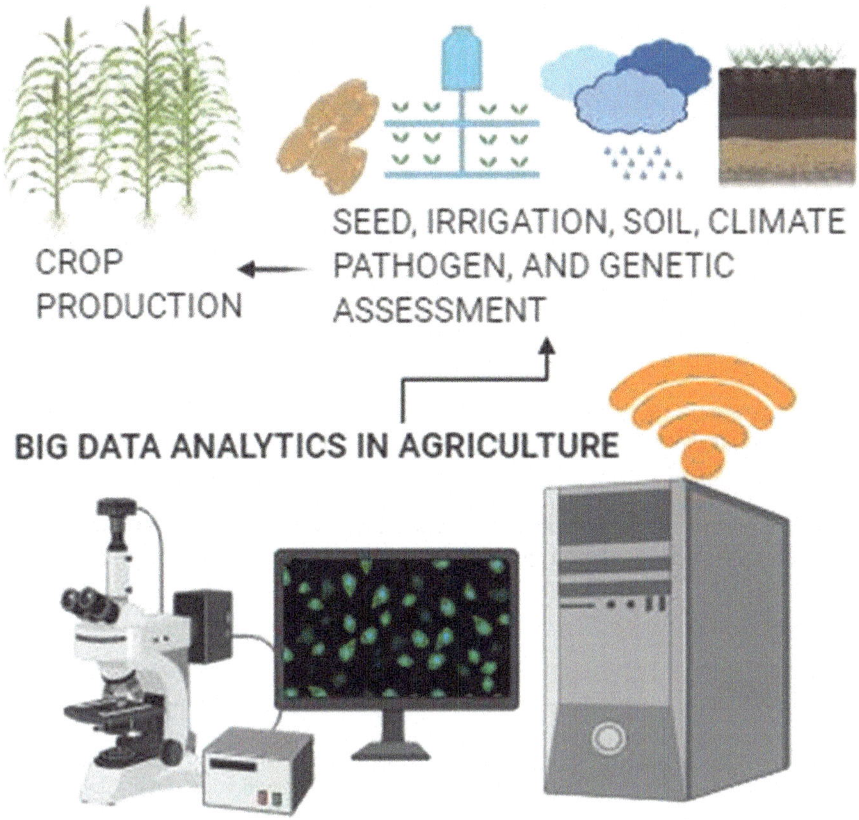

Fig. 4 Big data analytics in Agriculture: Major applications of big data management in Agriculture sector

crafting successful plan of actions to achieve sustainable development [28]. Sustainable development depends on sustainable system (Fig. 5) that shows interconnection between various research institutions, farms, and data repository sites to ensure lab to land transfer of information effectively.

Conclusion

Big data in agriculture is a term that describes all datasets, information regarding agriculture issues. It has various aspects: as it not only include transcriptomic-proteomic datasets, but also agribusiness and production databases. The analysis of enormous information is a challenging task, so different servers were designed for this purpose. These servers are selective in nature and provide ease of access to

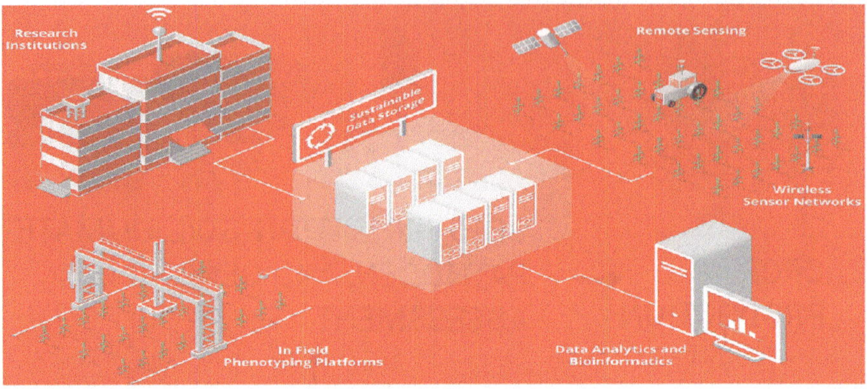

Fig. 5 Sustainable system: Data repository interconnection with remote sensing, research institutes, phenotyping platforms, data analytics

scientific groups. The only objective is the improvement of farmer's life and to maintain food security among individuals.

References

1. Bansal A, Srivastava PA. Transcriptomics to metabolomics: a network perspective for big data. In: Biotechnology: concepts, methodologies, tools, and applications: IGI Global; 2019. p. 361–79.
2. Yamanishi Y, Tabei Y, Kotera M. Statistical machine learning for agriculture and human health care based on biomedical big data. In: Forum "Math-for-Industry". Singapore: Springer; 2016. p. 111–23.
3. Kanaya S, Altaf-Ul-Amin M, Kiboi SK, Afendi FM. Big data and network biology 2015. BioMed Res Int. 2015;2015
4. Prabha R, Verma MK, Singh DP. Bioinformatics in agriculture: translating alphabets for transformation in the field. In: Plant bioinformatics. Cham: Springer; 2017. p. 197–214.
5. Muthuramalingam P, Krishnan SR, Pothiraj R, Ramesh M. Global transcriptome analysis of combined abiotic stress signaling genes unravels key players in Oryza sativa L.: an in silico approach. Front Plant Sci. 2017;8:759.
6. Lohse M, Nagel A, Herter T, May P, Schroda M, Zrenner R, et al. Mercator: a fast and simple web server for genome scale functional annotation of plant sequence data. Plant Cell Environ. 2014;37(5):1250–8.
7. Lee T, Oh T, Yang S, Shin J, Hwang S, Kim CY, et al. RiceNet v2: an improved network prioritization server for rice genes. Nucleic Acids Res. 2015;43(W1):W122–7.
8. Darwish O, Rowland LJ, Alkharouf NW. BBGD454: a database for transcriptome analysis of blueberry using 454 sequences. Bioinformation. 2013;9(17):883.
9. Xu ZS, Tan HW, Wang F, Hou XL, Xiong AS. CarrotDB: a genomic and transcriptomic database for carrot. Database. 2014;2014
10. Kim HJ, Baek KH, Lee SW, Kim J, Lee BW, Cho HS, et al. Pepper EST database: comprehensive in silico tool for analyzing the chili pepper (Capsicum annuum) transcriptome. BMC Plant Biol. 2008;8(1):101.

11. Zhai R, Feng Y, Wang H, Zhan X, Shen X, Wu W, et al. Transcriptome analysis of rice root heterosis by RNA-Seq. BMC Genomics. 2013;14(1):19.
12. Hirano H, Islam N, Kawasaki H. Technical aspects of functional proteomics in plants. Phytochemistry. 2004;65(11):1487–98.
13. Kuriakose SV, Pushker R, Hyde EM. Data-driven decisions for accelerated plant breeding. In: Accelerated plant breeding, vol. 1. Cham: Springer; 2020. p. 89–119.
14. Weckwerth W, Ghatak A, Bellaire A, Chaturvedi P, Varshney RK. PANOMICS meets germplasm. Plant Biotechnol J. 2020;
15. Das V, Jain S. Genetic algorithm to find most optimum growing technique for multiple cropping using big data. In: Emerging technologies for agriculture and environment. Singapore: Springer; 2020. p. 77–94.
16. Ercolano MR, Andolfo G, Frusciante L. Informatic tools and platforms for enhancing plant R-gene discovery process. In: Applied plant biotechnology for improving resistance to biotic stress: Academic Press; 2020. p. 121–35.
17. Song S, Zhang Z. Database resources in BIG data center: submission, archiving, and integration of big data in plant science. Mol Plant. 2019;12(3):279–81.
18. McCarthy FM, Wang N, Magee GB, Nanduri B, Lawrence ML, Camon EB, et al. AgBase: a functional genomics resource for agriculture. BMC Genomics. 2006;7(1):229.
19. Naithani S, Gupta P, Preece J, D'Eustachio P, Elser JL, Garg P, et al. Plant Reactome: a knowledgebase and resource for comparative pathway analysis. Nucleic Acids Res. 2020;48 (D1):D1093–103.
20. Gui S, Yang L, Li J, Luo J, Xu X, Yuan J, et al. ZEAMAP, a comprehensive database adapted to the maize multi-omics era. bioRxiv. 2020;
21. Zhou J, Li D, Wang G, Wang F, Kunjal M, Joldersma D, Liu Z. Application and future perspective of CRISPR/Cas9 genome editing in fruit crops. J Integr Plant Biol. 2020;62 (3):269–86.
22. Price EJ, Drapal M, Perez-Fons L, Amah D, Bhattacharjee R, Heider B, et al. Metabolite database for root, tuber, and banana crops to facilitate modern breeding in understudied crops. Plant J. 2020;101(6):1258–68.
23. Martins RC, Magalhães S, Jorge P, Barroso T, Santos F. Metbots: metabolomics robots for precision viticulture. In: EPIA conference on artificial intelligence. Cham: Springer; 2019. p. 156–66.
24. Bhatia A, Sarma SJ, Lei Z, Sumner LW. UHPLC-QTOF-MS/MS-SPE-NMR: a solution to the metabolomics grand challenge of higher-throughput, confident metabolite identifications. In: NMR-based metabolomics. New York: Humana; 2019. p. 113–33.
25. Ma C, Zhang HH, Wang X. Machine learning for Big Data analytics in plants. Trends Plant Sci. 2014;19(12):798–808.
26. Atzberger C. Advances in remote sensing of agriculture: context description, existing operational monitoring systems and major information needs. Remote Sens. 2013;5(2):949–81.
27. Verma S, Bhatia A, Chug A, Singh AP. Recent advancements in multimedia big data computing for IoT applications in precision agriculture: opportunities, issues, and challenges. In: Multimedia big data computing for IoT applications. Singapore: Springer; 2020. p. 391–416.
28. PS MG, Chintala BR. Big data challenges and opportunities in agriculture. Int J Agric Environ Inf Syst (IJAEIS). 2020;11(1):48–66.

The Distinction of Omics in Amelioration of Food Crops Nutritional Value

Bhupender Singh, Dibyalochan Mohanty, Vasudha Bakshi, Ranjit Singh Gujjar, and Atul Kumar Upadhyay

Introduction

Nutritional deficiency is one of the global concerns mainly in undeveloped and developing countries. Uptake of insufficient nutrient than required for normal growth and development results in a decline of health and subsequently compromise with the immunity of the individual causing more susceptible to numerous diseases. Ultimately reflects in a decline of the GDP (Gross Domestic Product), which is around 11% in Africa and Asia as most of the countries in these two continents are undeveloped or developing. An estimate has found that around 200 million of the world's population is nutritionally deficient and among those children are the largest in number. Moreover, approximately 45% of deceased kids below five years are because of nutritional deficiency. In India, 15.2% of the population is suffering from nutritional scarcity [1].

Unequal level of required nutrients in the food corresponds to the conditions related to the nutrient deficiency and commonly termed as "hidden hunger". One of the main tactics employed to overcome the situation is the assorted diet consumption, which includes vegetable, cereal, animal protein and fruits. However, this approach is not much feasible in developing countries because of the population with small earning [2]. Another strategy is to go for the nutrient fortification in the

B. Singh
Department of Biotechnology, Lovely Professional University, Jalandhar, India

D. Mohanty · V. Bakshi
Department of Pharmaceutics, Anurag University, Hyderabad, India

R. S. Gujjar
Division of Crop Improvement, Indian Institute of Sugarcane Research, Lucknow, India

A. K. Upadhyay (✉)
Department of Biotechnology, Thapar Institute of Engineering & Technology, Patiala, India
e-mail: atul.upadhyay@thapar.edu

A. K. Upadhyay et al. (eds.), *Bioinformatics for agriculture: High-throughput approaches*, https://doi.org/10.1007/978-981-33-4791-5_5

food or tablets. The major problem with the approach is its vigorous dispensation, lack of standard substructures in developing countries and user obedience. Last but not least the most efficient and flexible approach to respond nutrition deficiency is "Biofortification". The term refers to the implication of the genetic engineering manipulations to achieve nutritionally enriched crops [3].

Genetic engineering of the plants was initiated during the very first days of the start of agriculture (~8000–10,000 years back). Discerning breeding was brought into the picture during the early ages of agriculture to obtain superior plants. The practice of keeping the seeds of healthier plants and sow it during the upcoming season was carried out. Characters, viz. pathogen resistance, increased harvest, enhanced growth rate and bigger size of the fruit/vegetable were considered during the early ages of agriculture [4].

At present, the face of genetic engineering has been modulated by the influence of omics approaches. The advent of omics has tremendously helped in our knowledge to identify the useful and unwanted characters/traits and thus enhance the yield concerning the agricultural implications. Genomic, transcriptomic, proteomic and metabolomic studies have cemented the way to carry out efficient and productive plant breeding strategies to acquire disease/stress-resilient crops with high nutritional value and yield [5]. Omics allows understanding the expression of a respective trait as the intricate connection among genes, proteins and metabolites by referring to the system biology approaches. Systems biology is a multidisciplinary domain mainly involving chemistry, bioinformatics and computational biology to understand the individual cell/metabolite biology and its complex interactions as a part of the system [6].

Omics has been related to vast areas including but not limited to genomics, transcriptomics, proteomics and metabolomics. As discussed above the use of omics served as an elixir for agriculture, keeping in mind the population of individuals suffering from nutrient deficiency. In the upcoming part, we will have a brief discussion on multifaceted omics.

Genomics

Conventionally, use of the term omics came into the picture after the existence of genomics. The genome of a plant serves as a blueprint for its breeding history. GenBank maintained by NCBI (National Center for Biotechnology Information) stores the annotated DNA sequences from an array of genes and genomes and is available to the scientific community [7]. Genomics provides a platform to carry out molecular level crop biotechnology, breeding and assortment based on the molecular markers and thus speeding up the release of genetically altered productive crops. Apart from this, omics has licensed to produce industrially and pharmaceutically relevant compounds. Gene expression analysis reveals the target gene corresponding to the traits, this data can be utilized to achieve better-quality plant/crops [8].

Transcriptomics

The product of the transcription (devoid of introns), i.e. RNA is referred to as transcriptome, out of which mRNA (messenger RNA) codes for the protein. Synonymously it is also called as expression profiling as it helps in evaluating the expression level of a transcript respective of the gene. Transcriptome studies reveal the impact of inner and outer features on gene expression level. Transcriptome analysis coupled with NGS (Next-Generation Sequencing) technology enhances our ability to gain useful insights respective of functional components of a genome [9].

Proteomics

The expression of a particular gene results in the translation of protein accountable for various cellular mechanisms. The proteome helps in infer the expression of the messenger transcripts. Thousands of the proteins in the plants account for the shape, crop yield, taste and nutritional value. It is possible to identify the expression level of the proteins, which are involved during the pathogenesis or at any kind of stress conditions via protein expression profiling. Comparative proteome analysis helps in elucidation molecular process responsible for pathogen vulnerability or resistance. This can further be employed to raise pathogen-resistant traits of the crop [10].

Metabolomics

Metabolomics refers to the analysis of the chemical processes, which acts as a connection between genotype and phenotype. It helps to infer whether the translated protein is metabolically active or not and to identify the role of the active metabolites. The activity of the metabolite is influenced by internal and external conditions. Analysis of the metabolic interaction network helps in understanding the changes brought up by the various stress conditions and to understand the basic system biology. This can further be implemented to raise the improved crop traits. The metabolic analysis gives an immediate state of the cell at certain instances such as during pathogenesis, fruit ripening and finding metabolite responsible for the flavour and odour. Detection of the metabolite change during various instances can help in raising desirable crop traits. Metabolite profiling can also aid in inferring the mode of action of various pesticides, which serve as the basis for formulating novel pesticide entities [11] (Fig. 1).

In the upcoming part, we will have insight on omics strategies involved to ameliorate the crops nutritional value.

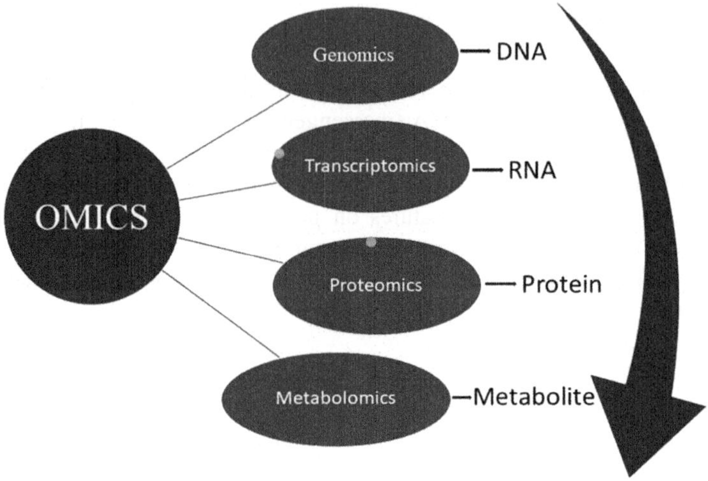

Fig. 1 General representation of the various omics approaches

Omics Intervention in Sugarcane

Sugarcane is a vital source for extraction of the sugar and globally around 80% of the sugar is processed from this plant. Moreover, the sugarcane is also utilized in the production of bioethanol and Brazil utilizes approximately 50% of its sugarcane for the production of ethanol. The omics intervention helps to comprehend the relationship between genomic composition, genes, proteins and metabolites and all of this has been processed by the synergy of bioinformatics and molecular biology techniques [12].

The genome of *S. officinarum* and *S. spontaneum* is 930 and 750 megabase pairs, respectively, [13]. These two species are being utilized as a backbone for the development of new cultivars, differs not only in their genome size but also in chromosome numbers because of transposable elements and chromosome reshuffling. Despite the intricate genome composition of the sugarcane, Genome-Wide Association Studies (GWAS) has paved the way to recognize marker-trait associations (MTAs) and ultimately help the breeders to opt superior genotypes to achieve successful breeding [12]. With the help of GWAS, the locus in genome responsible for resistance to yellow leaf virus [14] and orange rust disease [14] were identified in sugarcane and 23 MTAs were identified in sugarcane which corresponds to phenotypes, namely cane produce, solvable content and weight, height and number of stalks [15].

The transcriptomic analysis helps in uncovering all the relevant information related to a gene through various bioinformatic approaches and other experimental techniques. The Brazilian sugarcane EST (Expressed Sequenced Tags) database has obtained the data from 26 cDNA libraries of various sugarcane cultivars (from Brazil) and processed to constitute 238,000 ESTs. Further, these ESTs are catalogued into 43,141 exclusive transcripts which comprise 26,803 contigs. Apart from this, sugarcane gene index 3.0 consists of 282,683 ESTs with 499 cDNA having 121,342 unigenes [16]. Red rot in sugarcane is caused by *Colletotrichum falcatum*. The differential expression analysis revealed that 24 and 15 transcription factors after the fungal encounter and after systemic acquired resistance were found to express differentially, respectively. This analysis suggested that premature regulation of transcription factors may contribute to disease progression or acquiring resistance against the pathogen [17]. The pathogenesis of brown rust (*Puccinia melanocephala*) is being encountered by regulation of 11 (out of 217) resistance genes in sugarcane [18]. Transcriptome analysis of two different cultivars which were drought-tolerant (SP81-3250) and drought-vulnerable (RB855453) was carried out under various drought scenarios with the help of next-generation sequencing apparatus (HiScanSQ and HiSeq 2500) revealed that drought-tolerant cultivar was regulating SIZ2, ascorbate peroxidase and MYB which responds to combat abiotic stress. On the other hand, in RB855453 various kinases (RLK, bHLH, ACC oxidase and others) responsible for introducing stress were found to be regulated [19]. Besides, transcriptome analysis of sugarcane under potassium stressed conditions revealed 4153 differentially expressed genes and it was suggested that genes involved in signalling pathways (Ca^+ and ethylene signalling), oxidation stress, kinases, transporters and transcription factors are important to combat Potassium stress in sugarcane [20]. Similarly, transcriptome analysis revealed that MYB transcription factor gene family was differentially expressed principally in nitrogen-stress lenient cultivars ROC22 of sugarcane [21].

Somatic embryogenesis is a vital biotechnological intervention having major implication in micropropagation and breeding of sugarcane. Impact of the putrescine on the sugarcane was assessed with the help of proteome profiling and proteins, namely heat shock proteins, arabinogalactan, peroxidases, glutathione-S-transferases and late embryogenesis profuse proteins combat the putrescine mediated stress and helps in the generation of somatic embryo throughout development treatment processes [12].

Metabolomics help to establish the relationship between the traits and associated physiological variations to comprehend plant system. The metabolomic profiling of sugarcane is at its initial phase due to the difficulty faced in finding molecular markers corresponding to vital phenotypes and thus finding stable biologically active molecules. In sugarcane metabolomics, principally carbohydrates (raffinose, fructose, glucose, sucrose and inositol) are distinguished with the help of Gas Chromatography-Mass Spectroscopy analysis [22]. Figure 2 represents the identified sugarcane metabolites along with their associated functionalities [12].

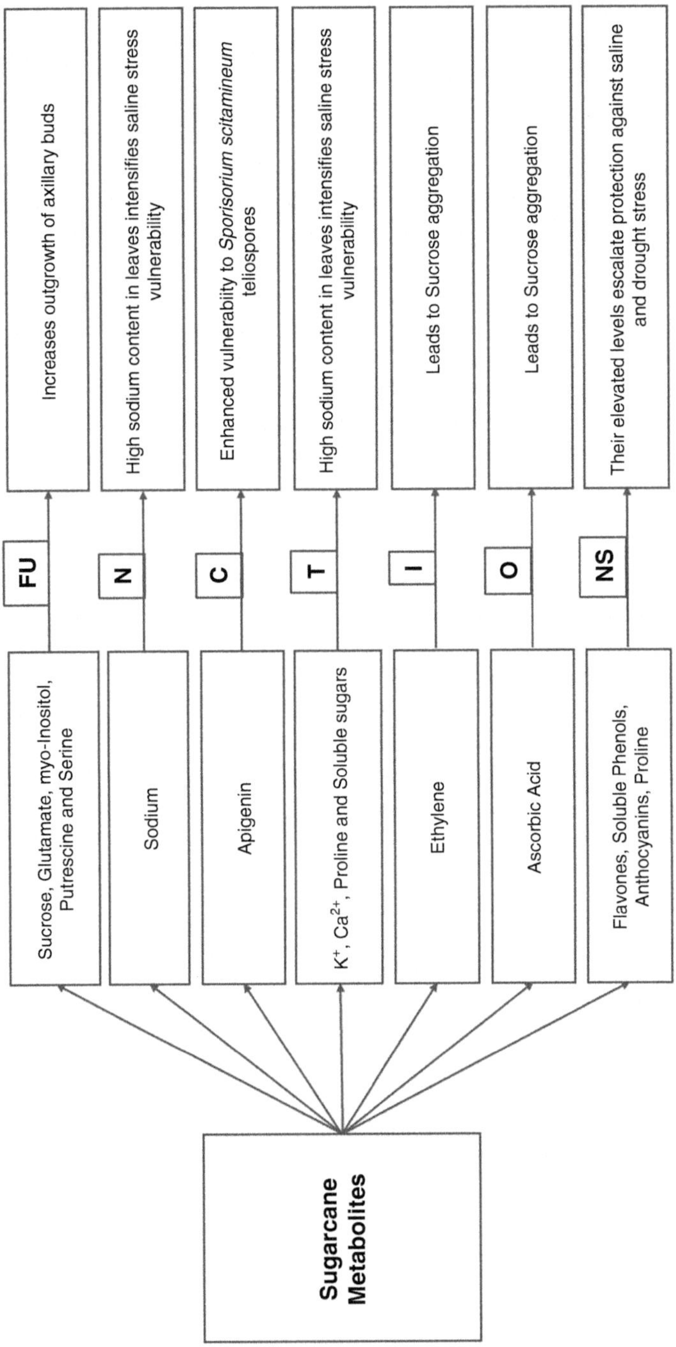

Fig. 2 Various sugarcane metabolites and their associated biological functions

Omics Intervention in Common Bean

Legumes are well known for their high protein, fibre, vitamin and mineral content at relatively less spending, as compared to the meat products and thus serve as an important food source in developing nations. Globally legumes contribute around 27% of fundamental crop produces. Common bean is among one of the legumes enriched with miscellaneous nutritional constituents like high protein content, adequate carbohydrates and little fat measures. Common bean is enriched with various phytochemicals which helps to combat multiple associated conditions, namely obesity, cardiovascular, blood glucose and colon malignancy. Considering the aforementioned properties, it is distinguished as "grain of hope" in economically unstable nations. Adverse environmental conditions resulting in biotic and abiotic stress greatly influence the common bean yield. To overcome such conditions, with the help of biotechnological implications new cultivars are being produced which are resistant to the adverse conditions and thus does not compromise with crop yield. One of the major difficulties faced by the breeders is to identify the multiple genes which account for a particular phenotype. Omics interventions have received major attention of the researchers to comprehend the physiology of the plants under various conditions and thus translate this information to produce genetically superior cultivars [23].

The germplasm of the various *Phaseolus* species is accessible at CIAT (Centro Internacional de Agricultura Tropical), Colombia, National Plant Germplasm System, USA (https://npgsweb.ars-grin.gov/gringlobal/search.aspx) and NBPGR (National Bureau of Plant Genetic Resources), India. For common bean, a sum of 102,738 entries is available in publicly accessible databases worldwide (http://legumecrops.wildsoydb.org) [23].

The sequencing associated annotation of the common bean can be accessed at https://phytozome.jgi.doe.gov/pz/portal.html#!info?alias=Org_Pvulgaris. The sequencing data were generated by using Pacific Biosciences sequencer and data assembly was carried out by FALCON (bioinformatics.ua.pt/software/falcon). The genome size of common bean is around 587 Mb out of which 478 scaffolds contain 537.2 Mb genome and 1044 contigs cover 521.6 Mb of the genome [24]. The availability of genome sequence has paved the way to distinguish molecular features amid numerous unfavourable conditions. Further, the transcriptome analysis reveals the specific transcripts which are being regulated under various stress environments, however, as the transcripts (after translation) undergo PTM (Post Translational Modifications) which erstwhile hinders the transcript to represent the ultimate function of associated genes. On the other hand, proteomics-based analysis is the gold standard to annotate a genome but it is not possible without implicating the information of genomics and transcriptomics [23]. The following representation highlights the differentially expressed proteins under numerous stress conditions and their associated functions to combat the same [23] (Fig. 3).

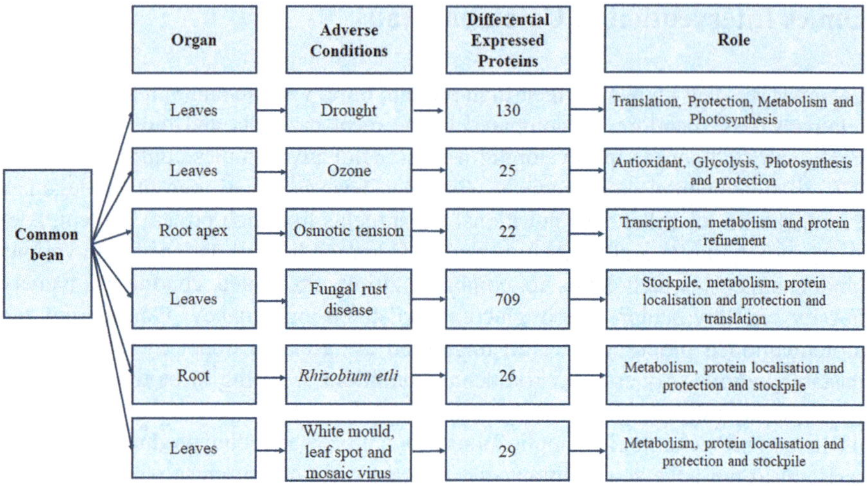

Fig. 3 Role of differentially expressed proteins under various adverse conditions

Omics Intervention in Finger Millet

Finger millet comes under the Poaceae family and *Eleusine* genus. It has a high quantity of protein and minerals as equated to sorghum, wheat and rice. Moreover, it contains around 0.34% of the calcium in the seeds which is relatively high as compared to other cereals whose calcium content ranges from 0.01 to 0.06%. Finger millet is also well known for its high fibre, amino acids, iron content and is devoid of gluten [25].

The genome of the finger millet is 1593 Mb [26]. In India, NBPGR (National Bureau of Plant Genetic Resources) and ICRISAT (International Crop Research Institute for the Semi-Arid Tropics) has 10,507 and 5957 number of germplasms stockpiled for finger millet. The whole-genome sequence for the finger millet cultivar ML-365, which is insensitive to blast infection and drought, was processed by next-generation sequencers, namely Illumina and SOLiD. Total 525,759 scaffolds were obtained with a mean scaffold length of 2275 base pairs [27]. Following transcriptomics, 53,300 (sufficient water employed) and 100,046 (kept water-deficient) genes were identified for ML-365 plants kept under different conditions. Differentially expressed genes investigation exposed that 12,893 and 2267 genes were explicit to plants which were sufficiently watered and water-deprived, respectively. Moreover, 111,096 genes were common in plants raised in the aforementioned environments. Further anticipation revealed that 11,125 genes were having correspondence with 56 transcription factor families [27]. Apart from this, 330 genes were found to be differentially expressed amid calcium transport and aggregation and 1766 genes were corresponding to provide resistance against pathogens [28]. Finger millet was genetically engineered by the incorporation of PIN and Chi11 genes differently which provide immunity against leaf blast pathogenesis

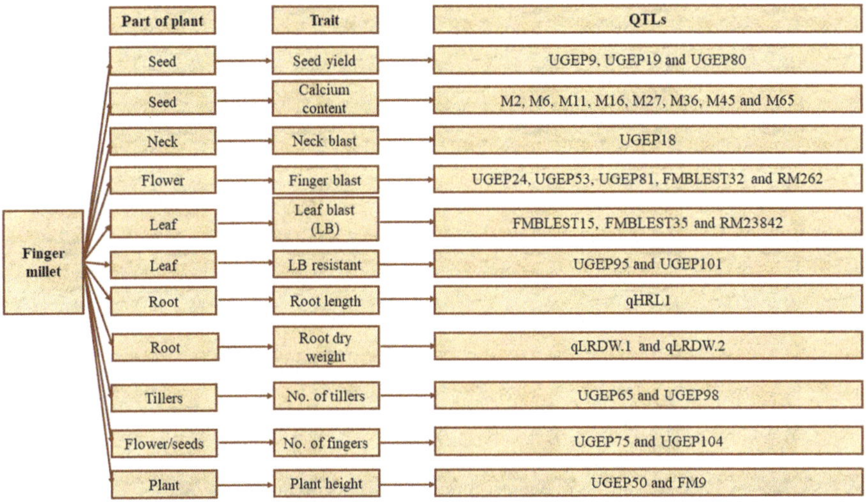

Fig. 4 Some of the vital QTLs identified in the finger millet

[29]. Molecular markers play a vital role in the determination of a particular phenotype for efficient breeding practices. Moreover, breeding practices assisted by molecular markers have an upper hand over traditional breeding approaches. Figure 4 represents some of the vital QTL (Quantitative Trait Locus) which are distinguished in finger millet employing molecular biology and omics strategies [25].

Omics Intervention in Rice

Rice is the principal staple food across 50% of the world's people and thus serves as a major energy source. Statistically, consumption of rice worldwide has elevated from 156 million tons in 1960 to 456 million tons in 2010 [30]. The nutritional composition of rice includes carbohydrate, protein, vitamin, fat, fibre, minerals and ash. It also helps to combat heart disease, high blood pressure, cancer, Alzheimer's and dysentery [31].

The IRGSP (International Rice Genome Sequencing Project) was formulated in 1998 which consists of eminent research groups from ten countries to obtain complete genome sequence of the rice (*O. sativa* ssp. *Japonica* cultivar Nipponbare). The sequenced genome was of 389 Mb with 95% coverage of the genome. Final processed genomic sequence was having less than one fault per 10,000 nucleotides. Further, 37,544 exon sequences were found having a gene density of 9.9 kb/gene. The analysis revealed that 2859 genes were exclusive of rice. The 35% of the analysed genomic region was corresponding to transposons. The prevalence of SNP (Single Nucleotide Polymorphism) ranges from 0.53 to 0.78%. Out of 37,544

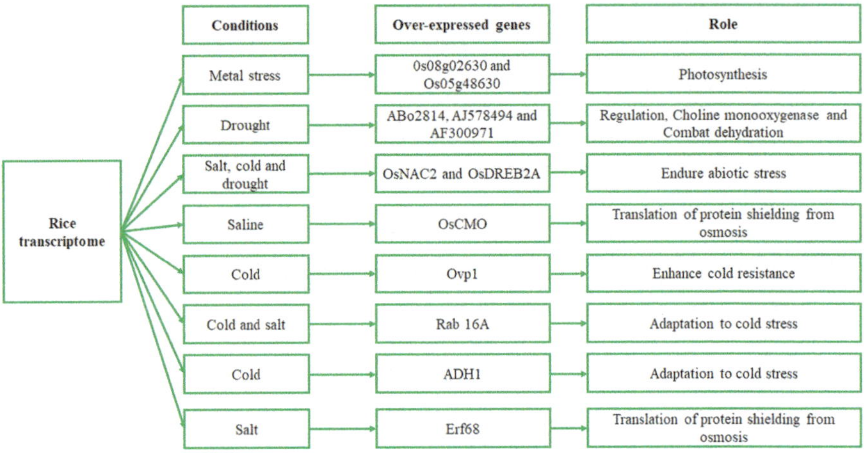

Fig. 5 Over-expressed genes in rice under numerous stress conditions

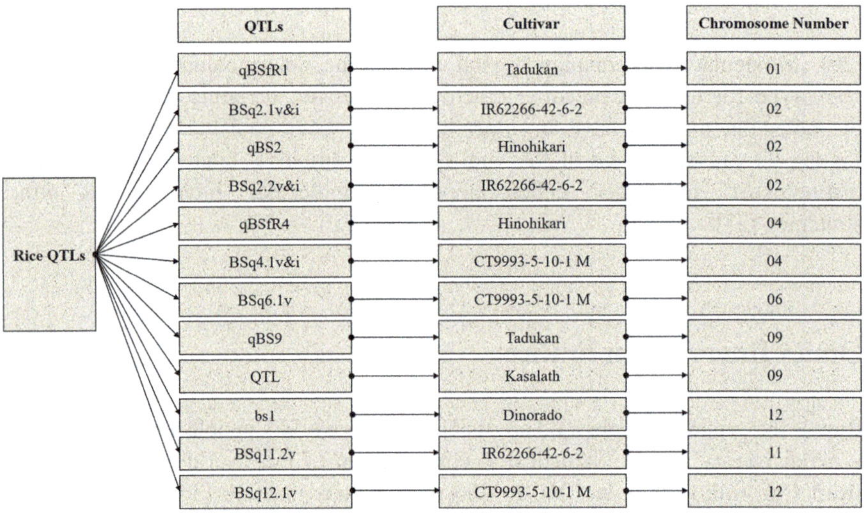

Fig. 6 QTLs associated with brown spot resistance in rice cultivars

non-transposon exon sequences recognized by FGENESH gene prediction tool, 17,016 were corresponding to 25,636 cDNAs. Moreover, around 61% that 22,840 genes were sharing a high degree of similarity with rice expressed sequence tag [32].

Transcriptome analysis of rice revealed the differentially expressed genes under numerous stress environments [33]. Figure 5 represents the genes which were found to be over-expressed under certain stress conditions [33].

In Figs. 6 and 7, identification of various QTLs associated with resistance against the pathogenesis of diseases, namely brown spot and bacterial grain rot which

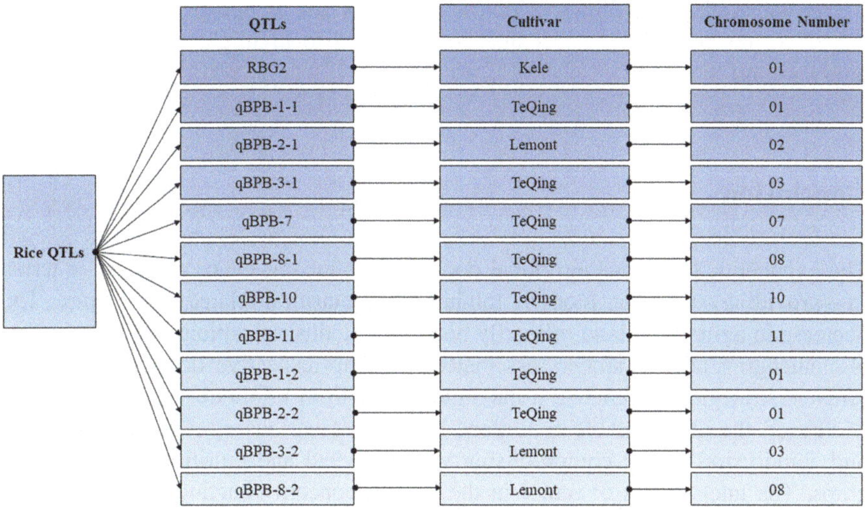

Fig. 7 QTLs associated with bacterial grain rot resistance in numerous rice cultivars

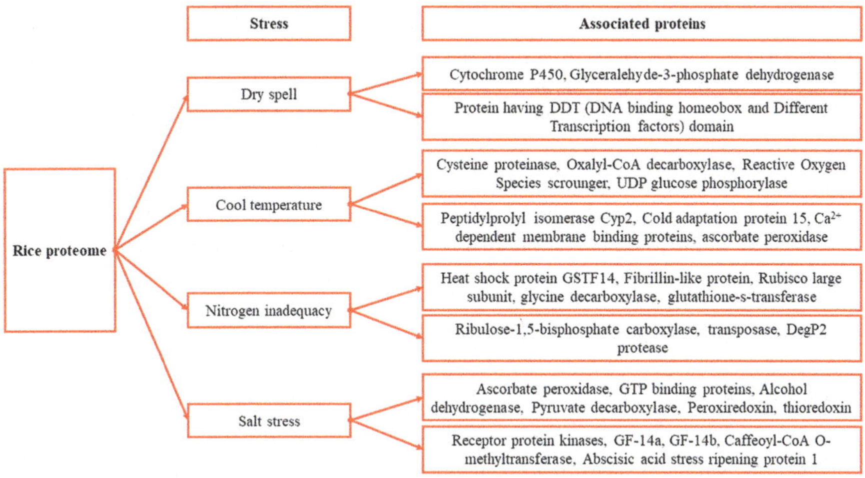

Fig. 8 Some vital proteins associated with rice under various stress conditions

compromises the rice crop yield and grain quality are presented [34]. The pathogen resistance QTLs identified in various rice cultivars are also associated.

Protein expression analysis is the gold standard to comprehend the regulation of proteins under various conditions and thus can provide leading outcomes to the breeders to achieve efficient cultivars. In Fig. 8 some of the vital proteins associated with various stress environments determined by various molecular biology and bioinformatics analysis are listed [33].

The intervention of omics in rice is enormous and is out of the scope of this chapter to discuss every aspect but we have tried to highlight some of the vital interventions.

Conclusion

The exponentially increasing global population is raising major concerns in terms of—providing sufficient food to minimize malnutrition-related disabilities. The decrease in agriculture land primarily because of industrialization and urbanization is challenging the researchers associated with the respective domain to develop efficient strategies to generate a maximum yield from minimum agricultural land. Moreover, the unfavourable conditions in terms of various crop pathogens, biotic and abiotic stresses are compromising with the yield and nutritional value of the crops. The intervention of omics in the present concern is acting like an elixir to quench the global food hunger. Omics analysis comprising of high-throughput techniques, namely next-generation sequencing, RNASeq expression analysis and set up of bioinformatics to process and analyse high-throughput data is helping to comprehend the biology of crops at multiple omics level, to extract the regulatory information of genes/transcripts/proteins/metabolites, which corresponds to combat various stress conditions. The information retrieved by these analyses is helping the researchers to produce genetically superior cultivars. We have attempted to show the impression of how effectively omics has been utilized in the identification of genomic locus associated with useful traits in various crops and further utilize this information to obtain genetically superior cultivars. More aggressive research and development in the associated domains will be vital to minimize/eliminate the global food hunger.

References

1. Yadava D, Hossain F, Mohapatra T. Nutritional security through crop biofortification in India: status & future prospects. Indian J Med Res. 2018;148:621. https://doi.org/10.4103/ijmr.IJMR_1893_18.
2. Gustafson D, Gutman A, Leet W, Drewnowski A, Fanzo J, Ingram J. Seven food system metrics of sustainable nutrition security. Sustain. 2016;8:196. https://doi.org/10.3390/su8030196.
3. Chadare FJ, Idohou R, Nago E, Affonfere M, Agossadou J, Fassinou TK, Kénou C, Honfo S, Azokpota P, Linnemann AR, Hounhouigan DJ. Conventional and food-to-food fortification: an appraisal of past practices and lessons learned. Food Sci Nutr. 2019;7:2781–95.
4. Boukid F, Folloni S, Sforza S, Vittadini E, Prandi B. Current trends in ancient grains-based foodstuffs: insights into nutritional aspects and technological applications. Compr Rev Food Sci Food Saf. 2018;17:123–36. https://doi.org/10.1111/1541-4337.12315.
5. Pathak RK, Baunthiyal M, Pandey D, Kumar A. Augmentation of crop productivity through interventions of omics technologies in India: challenges and opportunities. 3 Biotech. 2018;8:454. https://doi.org/10.1007/s13205-018-1473-y.

6. Hasin Y, Seldin M, Lusis A. Multi-omics approaches to disease. Genome Biol. 2017;18
7. Manzoni C, Kia DA, Vandrovcova J, Hardy J, Wood NW, Lewis PA, Ferrari R. Genome, transcriptome and proteome: the rise of omics data and their integration in biomedical sciences. Brief Bioinform. 2018;19:286–302. https://doi.org/10.1093/BIB/BBW114.
8. Kujur A, Saxena MS, Bajaj D, Laxmi PSK. Integrated genomics and molecular breeding approaches for dissecting the complex quantitative traits in crop plants. J Biosci. 2013;38:971–87.
9. Wolf JBW. Principles of transcriptome analysis and gene expression quantification: an RNA-seq tutorial. Mol Ecol Resour. 2013;13:559–72. https://doi.org/10.1111/1755-0998.12109.
10. Eldakak M, Milad SIM, Nawar AI, Rohila JS. Proteomics: a biotechnology tool for crop improvement. Front Plant Sci. 2013;4
11. Kumar R, Bohra A, Pandey AK, Pandey MK, Kumar A. Metabolomics for plant improvement: status and prospects. Front Plant Sci. 2017;8:1302.
12. Ali A, Khan M, Sharif R, Mujtaba M, Gao SJ. Sugarcane omics: an update on the current status of research and crop improvement. Plants. 2019;8:1–24. https://doi.org/10.3390/plants8090344.
13. Mustafa G, Joyia FA, Anwar S, Parvaiz A, Khan MS. Biotechnological interventions for the improvement of sugarcane crop and sugar production. In: Sugarcane - technology and research: InTech; 2018.
14. Yang X, Sood S, Luo Z, Todd J, Wang J. Genome-wide association studies identified resistance loci to orange rust and yellow leaf virus diseases in sugarcane (Saccharum spp.). Phytopathology. 2019;109:623–31. https://doi.org/10.1094/PHYTO-08-18-0282-R.
15. Barreto FZ, Bachega Feijó Rosa JR, Almeida Balsalobre TW, Pastina MM, Silva RR, Hoffmann HP, de Souza AP, Franco Garcia AA, Carneiro MS. A genome-wide association study identified loci for yield component traits in sugarcane (Saccharum spp.). PLoS One. 2019;14 https://doi.org/10.1371/journal.pone.0219843.
16. Xu S, Wang J, Shang H, Huang Y, Yao W, Chen B, Zhang M. Transcriptomic characterization and potential marker development of contrasting sugarcane cultivars. Sci Rep. 2018;8 https://doi.org/10.1038/s41598-018-19832-x.
17. Prasanth CN, Viswanathan R, Krishna N, Malathi P, Ramesh Sundar A, Tiwari T. Unraveling the genetic complexities in gene set of sugarcane red rot pathogen Colletotrichum falcatum through transcriptomic approach. Sugar Tech. 2017;19:604–15. https://doi.org/10.1007/s12355-017-0529-3.
18. Avellaneda MC, Parco AP, Hoy JW, Baisakh N. Putative resistance-associated genes induced in sugarcane in response to the brown rust fungus, Puccinia melanocephala and their use in genetic diversity analysis of Louisiana sugarcane clones. Plant Gene. 2018;14:20–8. https://doi.org/10.1016/j.plgene.2018.04.002.
19. da Silva MD, de Oliveira Silva RL, Ferreira Neto JRC, Benko-Iseppon AM, Kido EA. Genotype-dependent regulation of drought-responsive genes in tolerant and sensitive sugarcane cultivars. Gene. 2017;633:17–27. https://doi.org/10.1016/j.gene.2017.08.022.
20. Zeng Q, Ling Q, Fan L, Li Y, Hu F, Chen J, Huang Z, Deng H, Li Q, Qi Y. Transcriptome profiling of sugarcane roots in response to low potassium stress. PLoS One. 2015;10 https://doi.org/10.1371/journal.pone.0126306.
21. Yang Y, Gao S, Su Y, Lin Z, Guo J, Li M, Wang Z, Que Y, Xu L. Transcripts and low nitrogen tolerance: regulatory and metabolic pathways in sugarcane under low nitrogen stress. Environ Exp Bot. 2019;163:97–111. https://doi.org/10.1016/j.envexpbot.2019.04.010.
22. Naron DR, Collard FX, Tyhoda L, Görgens JF. Production of phenols from pyrolysis of sugarcane bagasse lignin: catalyst screening using thermogravimetric analysis – thermal desorption – gas chromatography – mass spectroscopy. J Anal Appl Pyrolysis. 2019;138:120–31. https://doi.org/10.1016/j.jaap.2018.12.015.

23. Zargar SM, Mahajan R, Nazir M, Nagar P, Kim ST, Rai V, Masi A, Ahmad SM, Shah RA, Ganai NA, Agrawal GK, Rakwal R. Common bean proteomics: present status and future strategies. J Proteomics. 2017;169:239–48. https://doi.org/10.1016/j.jprot.2017.03.019.
24. Schmutz J, McClean PE, Mamidi S, Wu GA, Cannon SB, Grimwood J, Jenkins J, Shu S, Song Q, Chavarro C, Torres-Torres M, Geffroy V, Moghaddam SM, Gao D, Abernathy B, Barry K, Blair M, Brick MA, Chovatia M, Gepts P, Goodstein DM, Gonzales M, Hellsten U, Hyten DL, Jia G, Kelly JD, Kudrna D, Lee R, Richard MMS, Miklas PN, Osorno JM, Rodrigues J, Thareau V, Urrea CA, Wang M, Yu Y, Zhang M, Wing RA, Cregan PB, Rokhsar DS, Jackson SA. A reference genome for common bean and genome-wide analysis of dual domestications. Nat Genet. 2014;46:707–13. https://doi.org/10.1038/ng.3008.
25. Antony Ceasar S, Maharajan T, Ajeesh Krishna TP, Ramakrishnan M, Victor Roch G, Satish L, Ignacimuthu S. Finger millet [Eleusine coracana (L.) Gaertn.] improvement: current status and future interventions of whole genome sequence. Front Plant Sci. 2018;9 https://doi.org/10.3389/fpls.2018.01054.
26. Goron TL, Raizada MN. Genetic diversity and genomic resources available for the small millet crops to accelerate a New Green Revolution. Front Plant Sci. 2015;6 https://doi.org/10.3389/fpls.2015.00157.
27. Hittalmani S, Mahesh HB, Shirke MD, Biradar H, Uday G, Aruna YR, Lohithaswa HC, Mohanrao A. Genome and Transcriptome sequence of Finger millet (Eleusine coracana (L.) Gaertn.) provides insights into drought tolerance and nutraceutical properties. BMC Genomics. 2017;18:465. https://doi.org/10.1186/s12864-017-3850-z.
28. Mirza N, Taj G, Arora S, Kumar A. Transcriptional expression analysis of genes involved in regulation of calcium translocation and storage in finger millet (Eleusine coracana L. Gartn.). Gene. 2014;550:171–9. https://doi.org/10.1016/j.gene.2014.08.005.
29. Ramakrishnan M, Antony Ceasar S, Duraipandiyan V, Vinod KK, Kalpana K, Al-Dhabi NA, Ignacimuthu S. Tracing QTLs for leaf blast resistance and agronomic performance of finger millet (Eleusine coracana (L.) Gaertn.) genotypes through association mapping and in silico comparative genomics analyses. PLoS One. 2016;11:e0159264. https://doi.org/10.1371/journal.pone.0159264.
30. Kaur B, Ranawana V, Henry J. The glycemic index of rice and rice products: a review, and table of GI values. Crit Rev Food Sci Nutr. 2016;56:215–36.
31. Verma DK. Nutritional value of rice and their importance. Indian Farmer's Dig. 2014;44:21.
32. Matsumoto T, Wu J, Kanamori H, Katayose Y, Fujisawa M, Namiki N, Mizuno H, Yamamoto K, Antonio BA, Baba T, Sakata K, Nagamura Y, Aoki H, Arikawa K, Arita K, Bito T, Chiden Y, Fujitsuka N, Fukunaka R, Hamada M, Harada C, Hayashi A, Hijishita S, Honda M, Hosokawa S, Ichikawa Y, Idonuma A, Iijima M, Ikeda M, Ikeno M, Ito K, Ito S, Ito T, Ito Y, Ito Y, Iwabuchi A, Kamiya K, Karasawa W, Kurita K, Katagiri S, Kikuta A, Kobayashi H, Kobayashi N, MacHita K, Maehara T, Masukawa M, Mizubayashi T, Mukai Y, Nagasaki H, Nagata Y, Naito S, Nakashima M, Nakama Y, Nakamichi Y, Nakamura M, Meguro A, Negishi M, Ohta I, Ohta T, Okamoto M, Ono N, Saji S, Sakaguchi M, Sakai K, Shibata M, Shimokawa T, Song J, Takazaki Y, Terasawa K, Tsugane M, Tsuji K, Ueda S, Waki K, Yamagata H, Yamamoto M, Yamamoto S, Yamane H, Yoshiki S, Yoshihara R, Yukawa K, Zhong H, Yano M, Sasaki T, Yuan Q, Ouyang S, Liu J, Jones KM, Gansberger K, Moffat K, Hill J, Bera J, Fadrosh D, Jin S, Johri S, Kim M, Overton L, Reardon M, Tsitrin T, Vuong H, Weaver B, Ciecko A, Tallon L, Jackson J, Pai G, Van Aken S, Utterback T, Reidmuller S, Feldblyum T, Hsiao J, Zismann V, Iobst S, De Vazeille AR, Buell CR, Ying K, Li Y, Lu T, Huang Y, Zhao Q, Feng Q, Zhang L, Zhu J, Weng Q, Mu J, Lu Y, Fan D, Liu Y, Guan J, Zhang Y, Yu S, Liu X, Zhang Y, Hong G, Han B, Choisne N, Demange N, Orjeda G, Samain S, Cattolico L, Pelletier E, Couloux A, Segurens B, Wincker P, D'Hont A, Scarpelli C, Weissenbach J, Salanoubat M, Quetier F, Yu Y, Kim HR, Rambo T, Currie J, Collura K, Luo M, Yang TJ, Ammiraju JSS, Engler F, Soderlund C, Wing RA, Palmer LE, De La Bastide M, Spiegel L, Nascimento L, Zutavern T, O'Shaughnessy A, Dike S, Dedhia N, Preston R, Balija V, McCombie WR, Chow TY, Chen HH, Chung MC, Chen

CS, Shaw JF, Wu HP, Hsiao KJ, Chao YT, Chu MK, Cheng CH, Hour AL, Lee PF, Lin SJ, Lin YC, Liou JY, Liu SM, Hsing YI, Raghuvanshi S, Mohanty A, Bharti AK, Gaur A, Gupta V, Kumar D, Ravi V, Vij S, Kapur A, Khurana P, Khurana P, Khurana JP, Tyagi AK, Gaikwad K, Singh A, Dalal V, Srivastava S, Dixit A, Pal AK, Ghazi IA, Yadav M, Pandit A, Bhargava A, Sureshbabu K, Batra K, Sharma TR, Mohapatra T, Singh NK, Messing J, Nelson AB, Fuks G, Kavchok S, Keizer G, Llaca ELV, Song R, Tanyolac B, Young S, Ho K, Hahn JH, Sangsakoo G, Vanavichit A, De Mattos LAT, Zimmer PD, Malone G, Dellagostin O, De Oliveira AC, Bevan M, Bancroft I, Minx P, Cordum H, Wilson R, Cheng Z, Jin W, Jiang J, Leong SA, Iwama H, Gojobori T, Itoh T, Niimura Y, Fujii Y, Habara T, Sakai H, Sato Y, Wilson G, Kumar K, McCouch S, Juretic N, Hoen D, Wright S, Bruskiewich R, Bureau T, Miyao A, Hirochika H, Nishikawa T, Kadowaki KI, Sugiura M, Burr B. The map-based sequence of the rice genome. Nature. 2005;436:793–800. https://doi.org/10.1038/nature03895.

33. Fahimirad S, Ghorbanpour M. Omics approaches in developing abiotic stress tolerance in rice (Oryza sativa L.): Elsevier Inc.; 2019.
34. Mizobuchi R, Fukuoka S, Tsushima S, Yano M, Sato H. QTLs for resistance to major rice diseases exacerbated by global warming: brown spot, bacterial seedling rot, and bacterial grain rot. Rice. 2016;9 https://doi.org/10.1186/s12284-016-0095-4.

Immunoinformatics in Plant–Fungal Disease Management

Sonika Nehra, Mahnoor Patel, Rekha Rani Das, and M. Amin-ul Mannan

Introduction

With the current rate of population growth, the world's population will be ~10 billion by 2050 [10]. Such a huge population will disturb global food security. With the limited resources, modest economic growths, dietary restriction it will further deteriorate the depleting natural resources. Agriculture is struggling to support the growing population food demands, addition of the plant disease and crop losses; adds to the catastrophe. It is estimated that plant pathogens, weeds, and animals together cause ~40% losses of global agricultural production [31]. Further toxins produced by postharvest pathogens also create serious health problems for consumers. Plant–fungal pathogens cause an enormous loss in quality and yield of crops, including fruits and other horticultural plants. Fungal diseases majorly affect five of the most important crops (wheat, rice, maize, soybean, and potato). It was predicted that if the loses are mitigated in can feed ~10% of world's population [9]. Decoding fungal pathogenesis will help in understanding the plant–fungal interaction and perhaps suggest the novel strategies for prevention, delay or inhibiting the fungal disease development and its control [3].

Immunoinformatics involves the application of computational methods to immunological problems. Prediction of B- and T-cell epitopes has been the focus of immunoinformatics. With the advent of next-generation sequencing (NGS) methods, an unprecedented wealth of information has become available that requires more-advanced immunoinformatics tools. Based on information from whole-genome sequencing, exome sequencing, and RNA sequencing much of the data needs to be analyzed and presented in the form of the database and datasets to the

S. Nehra · M. Patel · R. R. Das · M. A.-u. Mannan (✉)
Department of Molecular Biology and Genetic Engineering, School of Bioengineering and Biosciences, Lovely Professional University, Jalandhar-Delhi, Punjab, India
e-mail: mohammad.20597@lpu.co.in

A. K. Upadhyay et al. (eds.), *Bioinformatics for agriculture: High-throughput approaches*, https://doi.org/10.1007/978-981-33-4791-5_6

scientists. The chapter is an attempt to compile all the immunoinformatics tools which are associated with fungal–plant disease management.

Fungal pathogens of plants play a vital role in quantity, quality and profitability of the production. Farmers spend billions of dollars on disease management, often without adequate technical support, resulting in economic losses, environmental pollution, and harmful results. In addition, plant disease can devastate natural ecosystems, compounding environmental problems caused by habitat loss and poor land management [12]. Fungal pathogens of plants are responsible for causing various diseases of plants in horticulture and agriculture [11]. Cooperatively, phytopathogens have developed the mechanisms and various ways for attacking plants and looking for the entry and nutrients source for its own growth and development [18]. It affects negatively plant health, homeostasis, physiology and in some other cases cause systematic damage. Sexual dimorphism (asexual to sexual state) presents in plant fungus also allow them to overcome plant immune defense mechanism [1]. In this present chapter we have comprehended the various immunoinformatics tools, databases and *in silico* methods which can be used for the prediction of fungal diseases, plant–fungal interaction, virulence factors and its management thereof. We have also enlisted various fungal diseases, host factors, and plant defensive response and tools to predict the same. We envision the information provided here can be used by plant pathologist and experts involved in fungal disease managements.

Fungal–Plant Interaction

Based on the mode of infection fungal pathogens can be grouped as necrotrophs, biotrophs, and hemibiotrophs [17]. Biotrophs survive on the tissues of living organisms which cause an infection without killing its host. It uses appressorium for penetration and haustoria for deriving nutrients from the surrounding cells and also aids the secretion and translocation of effector molecules in the host plant. Biotrophs have a narrow host range, e.g. rust fungi and powdery mildew fungi [22]. Necrotrophs fungus infects the living host and eventually kills the host or region of the infestation. It completes its life cycle on the dead tissues. It uses hydrolytic enzymes including toxins for destroying plant cells. Hemibiotrophs also uses the similar mechanism like biotrophs which can cause infections and then killing of their host as necrotrophs [13]. Plants use its defense mechanisms to control the growth of pathogens.

Fungal Invasion Strategies

Phytopathogens are microorganisms that cause diseases in plants. These microorganisms are bacteria, viruses, fungi, nematodes, and protozoa. Plant–fungal pathogens use different strategies to attach and enter into the host [36]. As physical and

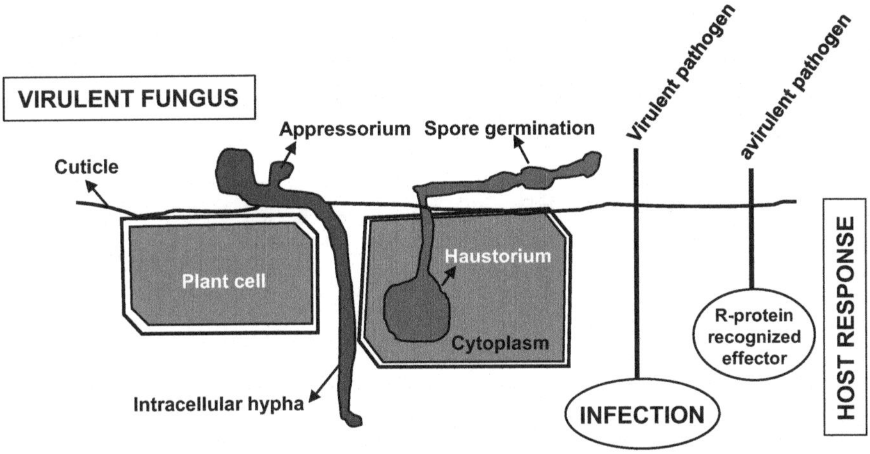

Fig. 1 Invasion of fungus into plant tissue. A virulent fungus breaches cuticle and cell wall and invade inside the plant cells or intracellular space using either germ tube or appressorium. Inside the cell a specialized structure called as haustorium derived the essential nutrients. As a defense strategy the basal defense mechanism include pathogen associated molecular patterns (PAMP) (innate immune response) or R-protein mediated defense activates the immune cells and make a virulent strain to avirulent

chemical barrier serve as first line of defense, penetrating the cell wall is an important step during host invasion. During host invasion some effector molecules are released by fungal phytopathogen. Some fungi use physical force generated by appresoria while others also use cell wall degrading enzymes (CWDEs) [29] (Fig. 1). During evolution, plants also established various innate and adaptive defense mechanisms against the fungal pathogens (Fig. 2). Pathogen-associated molecular patterns (PAMPs) and microbe-associated molecular patterns (MAMPs) are recognized via their pattern recognition receptors (PRRs) present on the cells of innate immune system. Fungal pathogens also need different strategies to overcome the plant defense mechanism, i.e. PAMP triggered immunity (PTI) inducible defense including effectors triggered immunity (ETI) defense [4]. Pathogens which become successful in disease can able to overcome ETI and PTI plant defense system. Some fungal pathogens have evolved mechanism to alter their cell wall to remain undetected from the host defense response. Resistance (R) genes help to identify and trigger immune response. Pathogens which involved avirulence genes (Avr) to overcome the R-resistance genes of the host plant cause serious infection. Some of the major fungal avirulence factors and its targets are enlisted in Table 1.

For the controlling and prevention of the plant–fungal diseases, fungicides have been used by the farmers for the management of damage the pathogenic fungi of plant. Plants have also developed its defense system for the prevention and elimination of pathogens. Pathogen recognition lead to accumulation of large amount of reactive oxygen species (ROS), deposition of callose in cell walls, generation of antimicrobial compounds, activation of defense related mitogen activating protein

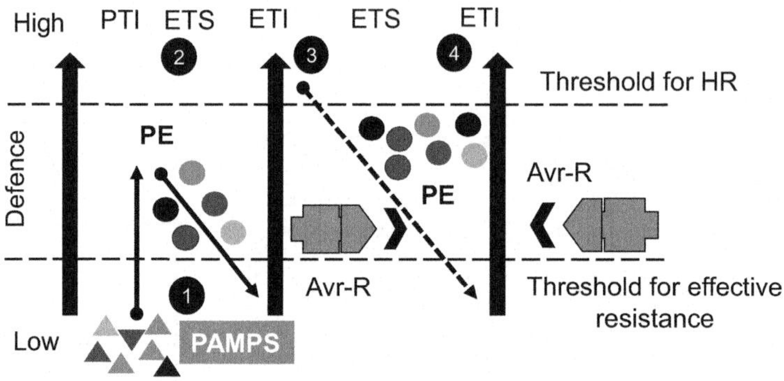

Fig. 2 Model of plant immune system during fungal pathogenesis. (1) Plant recognized PAMPS (Pathogen associated molecular patterns) using pathogen recognition receptor (PRR) which elicit PTI (PAMP triggered immunity). (2) At this stage two possibility can occur the invaded pathogen either induce ETS (Effectors triggered susceptibility) or interfere with PTI. (3) If the host proteins properly recognizes the pathogen it lead to ETI: Effectors triggered Immunity. (4) In cycle 4, natural selection lead to the formation of the new effectors through horizontal gene transfer replacing the old effectors, and plants generate new *Resistant (R)* genes to resist pathogens, resulting in ETI again. Abbreviations: HR: Hypersensitive cell death response; PE: Pathogen effectors proteins; PTI: PAMP triggered immunity; ETS: Effectors triggered susceptibility; ETI: Effectors triggered Immunity (Adapted from [32])

Table 1 Fungal virulence factors involved in host invasion

Fungal virulence factors	Targets	References
Appressoria	Cell wall (turgor pressure)	Ryder and Talbot [30]
Cell wall degrading enzymes	Cell wall	Kleemann et al. [21]
Carbohydrate-active enzymes (CAZymes, hydrolases, glycosyltransferases, lyases, esterases, and redox enzymes)	Cell wall and membrane	Lombard et al. [23]
Pectinases	Cell wall penetrance	Alghisi and Favaron [2]
Modification of cell wall composition	Accumulating of α-1,3-glucans, reduction of β-1,3- and β-1,6-glucans	Oliveira-Garcia and Deising [26]
Acidification (oxalic acid), alkalinization (Rapid alkalization factors, RALFs)	pH change at the site of infection	Masachis et al. [25] and Bolton et al. [6]
Cercosporin	Generation of ROS causes protein and DNA damage and lipid peroxidation	Birben et al. [5]
Host-selective toxins (HSTs)	T-toxin and PM-toxin, victorin, Necrotrophic effectors	Tsuge et al. [35], Gilbert and Wolpert [15] and Friesen and Faris [14]

kinases (MAPK), release of some enzymes such as glucanase and chitinase to degrade fungal cell wall [29]. Drawbacks which include resistance development and toxicity of environment which associated with the chemicals motivated the plant researchers and cultivators for investigating the possibilities.

Functional Genomics and Proteomics Tools

Immunoinformatics is evolving as a promising field as it provides convenient tools to connect immunology to computational methods. It helps to manage huge amount of data generated by various fields such as genomics, proteomics, and transcriptomics. Immunological techniques are used to study host immune response to pathogens. Though most of the study has been done on human immune response, particularly B- & T-cell epitopes, the following techniques are found to be useful to study plant response to a fungal pathogen.

Sequence Retrieving Tools

The analytical study of any sequence of DNA, RNA or protein to determine its structure, feature, and function is known as sequence retrieving. Fungal Genomics contains reviews on gene identification, comparative genomics, secreted proteins, and EST assembly. GnpIS (Genoplante Information System) is one such genomic retrieving tool. It is modular web-based integrative information system for genetic data and genomic information for plant and fungal pest [34]. Three primary sequence databases: GenBank (NCBI), the Nucleotide Sequence Database (EMBL), and the DNA Databank of Japan (DDBJ) are considered as storehouse of raw genomic data. Smith–Waterman algorithm (SSEARCH), FASTA, BLAST CLUSTALW, COBALT, MUSCLE are some other important tools in the diagnosis and treatment of fungal disease in plants [28]. ESTs (expressed sequence tags) and microarray analysis is done to study the genes of plant and fungal pathogens involved during infection. FGENESH is a gene finding software. The WoLF PSORT software is gene comparative software [16].

Proteomics Tools

Study of genome gives structural information but it is the protein which acts as functional molecule for the stored information in nucleotides. Proteomics means protein sequence of genome. Swiss-Prot, TrEMBL are tools for storage proteins. PROSITE, PRINTS, and BLOCKS are considered as secondary databases for protein sequences as they use information from primary databases Model Organism

Databases (MODs) are mentioned as high level clade specific databases (Journal and Entomology 2012). EST derived data is used to generate various proteomics databases such as dbEST at the National Centre for Biotechnology Information (NCBI), COGEME, Phytopathogenic Fungi, and Oomycete EST Database at Exeter University, UK [16].

Gene Ontology and Data Retrieval Software's

A large amount of data is provided by genomics and proteomics. Gene ontology refers to annotation of all genes and genes products and determining its functions. GO Consortium & PAMGO (plant-associated microbe gene ontology) are some databases helpful in this context. The first Web-based GO browser AmiGO has also been designed [8]. Some other key databases in plant–fungal interaction study are—FGI (Fungal Genome Initiative), IMG (Integrated Microbial Genomes), JGI (Joint Genomic Institute), PEDANT (Protein Extraction, Description and Analysis Tool) [57] at MIPS (Munich Information Center for Protein Sequences), and VMD (VBI Microbial Database) [59] at VBI (Virginia Bioinformatics Institute). NGS (next-generation sequencing) machines, such as Roche (454 Titanium), Illumina (Genome Analyzer II), and ABI (AB Solid) have increased the speed and accuracy of data retrieving process. Plant–fungal pathogen study has reached a new height by using a combination of GO and NGS [33].

MicroRNA and Transcriptomics Tools or Databases

Transcriptomics refers to study of coding as well as noncoding RNAs. Transcriptomics databases are available to help in identifying the mechanism of fungal pathogenesis. Micro-RNA like RNAs (milRNAs) are detected in fungus associated with plant diseases. OVERLAP software is commonly used to carry out studies related to microRNAs [19]. Databases such as RepBase, PASA, SNAP, BUSCO are used to predict RNA sequences.

Plant–Fungal Management

Fungi are one of the major threats to agricultural and horticultural industry. Phytopathogens have special mechanisms for entry and nutrition inside host tissues (Fig. 1). Fungal spores germinate on cell surface of appropriate host only under favorable conditions. Presence of moisture and nutrients increases the chances of spore germination. During unfavorable conditions, fungal spores can remain viable and dormant for many years. They germinate once the situations are favorable. Upon germination, appressorium is formed which lead to peg formation to facilitate entry inside host cell by penetrating the cell wall. Fungal hyphae are formed to obtain nutrients from host cell. Phytopathogens are able to flourish and reproduce successfully using nutrition from the host which negatively affects plant health and cause damage. Though crop plants have developed defense mechanism against fungal pathogens but the pathogens also seems to co evolve and have generated various resistance and escaping mechanism from host immune system. The various tools which can be used for fungal–plant interaction are mentioned below.

Adhesin Prediction Software's for Plant–Fungal Pathogens

As adhesins play an important role during fungal pathogenesis, developing tools for comparative genomic study to predict the adhesions involved and their mode of infection might be useful for designing antifungal compounds for these phytopathogens. Most of the adhesion study has been carried out on human beings but "FungalRV adhesin predictor" might be useful for plants as well. But more tools focusing particularly on plants are yet to be made [7].

Secondary Metabolites Prediction and Its Role in the Pathogenesis

Secondary metabolites are produced both in plants and fungal pathogens. Fungal pathogens mainly produce four types of secondary metabolites (SM), namely; polyketides, terpenoids, shikimic acid derived compounds, and non-ribosomal peptides. Plants produce phenolics, alkaloids, flavonoids, and terpenoids as secondary metabolites [27]. A special portal named Secondary Metabolite Bioinformatics Portal (SMBP) has been designed to study the interaction between both types of SMs to provide an insight into mechanism of fungal infection. Identifying secondary metabolite biosynthetic gene clusters (BGCs) is an important aspect in this approach. BAGEL, ClustScan, and NaPDoS are some important databases which uses data retrieved from BLAST & Hidden Markov Model software (HMMer) [38].

Plant and Fungal Secretomics

Secretomics refers to the global study of proteins secreted by any cell, tissue or organ. Secretomics might be an important tool in developing fungal resistance crop varieties as most of the effectors released by fungal pathogen and molecules secreted by plants as immune response during plant interaction with fungal interaction are secreted proteins [37]. Many secreted protein databases such as fungal Secretome KnowledgeBase (FunSecKB), orysPSSP, SignalIP, etc. have been designed to identify and compare *in silico* the various proteins and enzymes involved during plant–fungal pathogenesis [24].

Databases for Integrated Pest Management

Various pests like bacteria, fungi, nematodes, and insects cause a great loss to economic crops. Integrated pest management is (IPM) a system of practicing various methods to control pests to reduce economic crop damage [7]. Numerous pests of crop plants have been identified leading to need of computational methods to store and analyze various genomic and functional information related to pests and their host to provide a better strategy for developing control methods. Pathogen–host interactions database (PHI-base) and Global pest and disease database (GPDD) act as a catalogues for pathogenicity, virulence, and effector genes from fungal bacterial pathogens [20].

Conclusions

To reduce crop loss due to fungal diseases, fungicides either of chemical origin or of plant origin are used. As the chemical fungicides have many adverse effects on environment, organic fungicides are used more frequently. Combination of plant based antifungal is usually preferred than single one as different molecules interact with different metabolic pathways and reduce the risk of resistance development in the fungal pathogen. The tools described here suggest some of the avenues which can be deployed for understanding host–fungal interaction. It perhaps also opens the door for other areas to be explored more. Most of the research is animal oriented, fields related to development of effective, appropriate, and affordable antifungal organic compounds for plant–fungal diseases are yet to be explored. To ensure food security to growing population and provide some economic benefits to farmers, there is need to focus on ways of decreasing crop losses due to bacterial, fungal, and other pathogens. Integrated pest management is a good approach in this direction. Apart from chemical or organic compounds, biological agents occurring as natural fungal predators can also be used to control pest population. Developing new tools for early detection and prevention can increase the crop yield.

Conflict of the Interest
Authors declare there is no conflict of the interest.

Acknowledgments The lab funding from the Scientific and Engineering Research (SERB), Core Research Grant, file no. EMR/2017/002299, India is duly acknowledged.

References

1. Agrios GN. Plant pathogens and disease: general introduction. In: Encyclopedia of microbiology: Elsevier Inc.; 2009. p. 613–46. https://doi.org/10.1016/B978-012373944-5.00344-8.
2. Alghisi P, Favaron F. Pectin-degrading enzymes and plant-parasite interactions. Eur J Plant Pathol. 1995;101:365–75.
3. Almeida F, Rodrigues ML, Coelho C. The still underestimated problem of fungal diseases worldwide. Front Microbiol. 2019;10(Feb):214. https://doi.org/10.3389/fmicb.2019.00214.
4. Bellincampi D, Cervone F, Lionetti V. Plant cell wall dynamics and wall-related susceptibility in plant-pathogen interactions. Front Plant Sci. Frontiers Media S.A. 2014; https://doi.org/10.3389/fpls.2014.00228.
5. Birben E, Sahiner UM, Sackesen C, Erzurum S, Kalayci O. Oxidative stress and antioxidant defense. World Allergy Organ J. 2012;5:9–19.
6. Bolton MD, Thomma BP, Nelson BD. Sclerotinia sclerotiorum (Lib.) de Bary: biology and molecular traits of a cosmopolitan pathogen. Mol Plant Pathol. 2006;7:1–16.
7. Chaudhuri R, Ansari FA, Raghunandanan MV, Ramachandran S. FungalRV: adhesin prediction and immunoinformatics portal for human fungal pathogens. BMC Genomics. 2011;12 (1):192. https://doi.org/10.1186/1471-2164-12-192.
8. Clark JI, Brooksbank C, Lomax J. It's all GO for plant scientists. Plant Physiol. 2005;138 (3):1268–79. https://doi.org/10.1104/pp.104.058529.
9. FAO. The future of food and agriculture – trends and challenges. Rome: FAO; 2017.
10. FAO. The future of food and agriculture – alternative pathways to 2050. Rome: FAO; 2018. p. 224.
11. Fisher MC, Henk DA, Briggs CJ, Brownstein JS, Madoff LC, McCraw SL, Gurr SJ. Emerging fungal threats to animal, plant and ecosystem health. Nature. 2012; https://doi.org/10.1038/nature10947.
12. Fisher MC, Gow NAR, Gurr SJ. Tackling emerging fungal threats to animal health, food security and ecosystem resilience. Philos Trans R Soc B: Biol Sci. 2016;371(1709) https://doi.org/10.1098/rstb.2016.0332.
13. Fletcher J, Bender C, Budowle B, Cobb WT, Gold SE, Ishimaru CA, Luster D, Melcher U, Murch R, Schem H, Seem RC, Sherwood JL, Sobral BW, Tolin SA. Plant pathogen forensics: capabilities, needs, and recommendations. Microbiol Mol Biol Rev. 2006;70
14. Friesen TL, Faris JD. Characterization of the wheat-Stagonospora nodorum disease system: what is the molecular basis of this quantitative necrotrophic disease interaction? Can J Plant Pathol. 2010;32:20–8.
15. Gilbert BM, Wolpert TJ. Characterization of the LOV1- mediated, victorin-induced, cell-death response with virus-induced gene silencing. Mol Plant-Microbe Interact. 2013;26:903–17.
16. González-Fernández R, Prats E, Jorrín-Novo JV. Proteomics of plant pathogenic fungi. J Biomed Biotechnol. 2010;2010 https://doi.org/10.1155/2010/932527.
17. Horbach R, Navarro-Quesada AR, Knogge W, Deising HB. When and how to kill a plant cell: infection strategies of plant pathogenic fungi. J Plant Physiol. 2011;168
18. Jain A, Sarsaiya S, Wu Q, Lu Y, Shi J. A review of plant leaf fungal diseases and its environment speciation. Bioengineered. 2019;10(1):409–24. https://doi.org/10.1080/21655979.2019.1649520.
19. Jiang X, Qiao F, Long Y, Cong H, Sun H. MicroRNA-like RNAs in plant pathogenic fungus Fusarium oxysporum f. sp. niveum are involved in toxin gene expression fine tuning. 3 Biotech. 2017;7(5):1–12. https://doi.org/10.1007/s13205-017-0951-y.
20. Kavi KPB, Bandopadhyay R, Suravajhala P. Agricultural bioinformatics. Agric Bioinform. 2014;9788132218(January 2018):1–291. https://doi.org/10.1007/978-81-322-1880-7.
21. Kleemann J, Rincon-Rivera LJ, Takahara H, et al. Sequential delivery of host-induced virulence effectors by appressoria and intracellular hyphae of the phytopathogen Colletotrichum higginsianum. PLoS Pathog. 2012;8:e1002643.

22. Liang P, Liu S, Xu F, Jiang S, Yan J, He Q, Liu W, et al. Powdery mildews are characterized by contracted carbohydrate metabolism and diverse effectors to adapt to obligate biotrophic lifestyle. Front Microbiol. 2018;9(December) https://doi.org/10.3389/fmicb.2018.03160.
23. Lombard V, Golaconda Ramulu H, Drula E, Coutinho PM, Henrissat B. The carbohydrate-active enzymes database (CAZy) in 2013. Nucleic Acids Res. 2014;42:490–4955.
24. Lum G, Min XJ. FunSecKB: The Fungal Secretome KnowledgeBase. Database. 2011;2011:1–10. https://doi.org/10.1093/database/bar001.
25. Masachis S, Segorbe D, Turra D, et al. A fungal pathogen secretes plant alkalinizing peptides to increase infection. Nat Microbiol. 2016;1:16043.
26. Oliveira-Garcia E, Deising HB. Attenuation of PAMP-triggered immunity in maize requires down-regulation of the key b-1,6-glucan synthesis genes KRE5 and KRE6 in biotrophic hyphae of Colletotrichum graminicola. Plant J. 2016;87:355–75.
27. Pusztahelyi T, Holb IJ, Pócsi I. Secondary metabolites in fungus-plant interactions. Front Plant Sci. 2015;6(Aug):1–23. https://doi.org/10.3389/fpls.2015.00573.
28. Rao S, Nandineni MR. Genome sequencing and comparative genomics reveal a repertoire of putative pathogenicity genes in chilli anthracnose fungus Colletotrichum truncatum. PLoS ONE. 2017;12 https://doi.org/10.1371/journal.pone.0183567.
29. Rodriguez-Moreno L, Ebert MK, Bolton MD, Thomma BPHJ. Tools of the crook- infection strategies of fungal plant pathogens. Plant J. 2018;93(4):664–74. https://doi.org/10.1111/tpj.13810.
30. Ryder LS, Talbot NJ. Regulation of appressorium development in pathogenic fungi. Curr Opin Plant Biol. 2015;26:8–13.
31. Savary S, Ficke A, Aubertot J, et al. Crop losses due to diseases and their implications for global food production losses and food security. Food Sec. 2012;4:519–37. https://doi.org/10.1007/s12571-012-0200-5.
32. Shen Y, Liu N, Li C, Wang X, Xu X, Chen W, Xing G, Zheng W. The early response during the interaction of fungal phytopathogen and host plant. Open Biol. 2017;7(5) https://doi.org/10.1098/rsob.170057.
33. Soderlund C. Computational techniques for elucidating plant-pathogen interactions from large-scale experiments on fungi and oomycetes. Brief Bioinform. 2009;10(6):654–63. https://doi.org/10.1093/bib/bbp053.
34. Steinbach D, Alaux M, Amselem J, Choisne N, Durand S, Flores R, Keliet AO, et al. GnpIS: an information system to integrate genetic and genomic data from plants and fungi. Database. 2013;2013:1–9. https://doi.org/10.1093/database/bat058.
35. Tsuge T, Harimoto Y, Akimitsu K, Ohtani K, Kodama M, Akagi Y, Egusa M, Yamamoto M, Otani H. Host-selective toxins produced by the plant pathogenic fungus Alternaria alternata. FEMS Microbiol Rev. 2013;37:44–66.
36. Vincent D, Rafiqi M, Job D. The multiple facets of plant–fungal interactions revealed through plant and fungal secretomics. Front Plant Sci. 2020;10(January) https://doi.org/10.3389/fpls.2019.01626.
37. Vincent D. Secretomics of plant-fungus associations: more secrets to unravel. J Plant Biochem Physiol. 2013;01(05):1–2. https://doi.org/10.4172/2329-9029.1000e117.
38. Weber T, Kim HU. The secondary metabolite bioinformatics portal: computational tools to facilitate synthetic biology of secondary metabolite production. Synth Syst Biotechnol. 2016;1(2):69–79. https://doi.org/10.1016/j.synbio.2015.12.002.

Agri/Bioinformatics: Shaping Next-Generation Agriculture

Richa Mishra and Dhananjay K. Pandey

Agri/Bio Informatics

Agri-bioinformatics offers an integrative network of computational science, fundamental genetics, agricultural science, statistics, and omics technologies, including genomics, transcriptomics, proteomics, epigenomics, metabolomics with high potential to transform mainstream agricultural practices and agroeconomics [10, 40, 123]. Agri-bioinformatics components, including the omic toolbox and databases help process and interpret enormous biological information in multidimensional form by using advanced computing technology (Fig. 1). Such informations are highly useful in improving agronomic traits and managing biotic/abiotic stress in crop plants. Food safety for the growing population is the major challenge for decades to come. Today's high demand for crop improvement is for higher yields by integrating new traits into conventional crops along with effective stress tolerance mechanisms in these plants [10]. Bioinformatics tools enable the generation, collection, and interpretation of biological data on key factors that are responsible for better crop yield. In the last ten years, this integrative science has evolved rapidly and could be a powerful driving force for the agricultural sectors [40]. In the age of high-throughput technology, when we talk about data generation, most of the data generated is in the form of biological sequences that could be DNA, RNA or protein. The biological knowledge of living organisms, as per the central dogma, moves in a particular order and flows from DNA to RNA to protein [10, 123]. As a result of recent progress in sequencing technology, there has been a large increase in the amount of biological data available from various agricultural plant species. There are already more than 236 species of angiosperm sequenced [27]. Publicly accessible biological data are growing dramatically every day due to reduced sequencing costs, the availability of powerful computing

R. Mishra · D. K. Pandey (✉)
Department of FF21, University of California, San Diego, CA, USA

A. K. Upadhyay et al. (eds.), *Bioinformatics for agriculture: High-throughput approaches*, https://doi.org/10.1007/978-981-33-4791-5_7

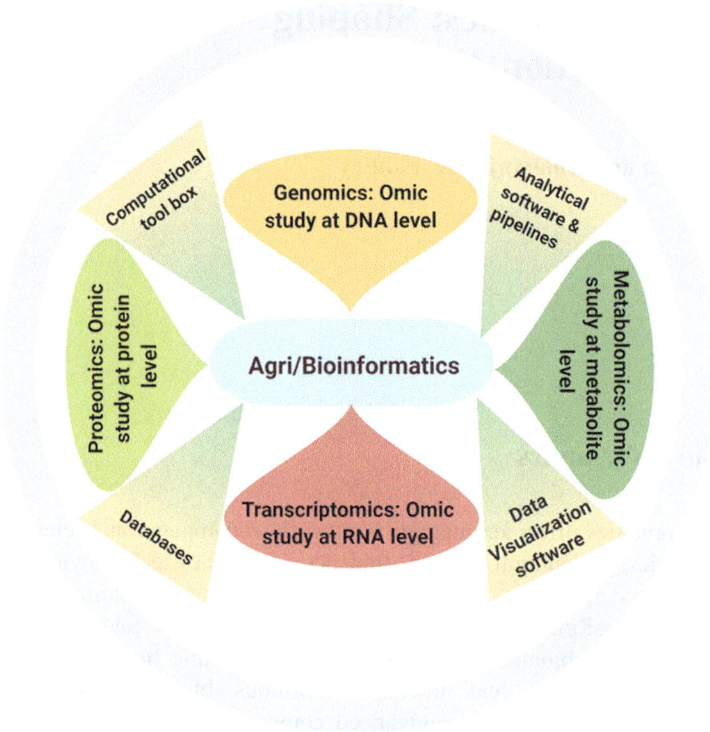

Fig. 1 Diagrammatic representation of the basic "omics" methods, databases, and computational tools used in Agri/Bioinformatics

resources, and data storage capabilities [10, 40, 123]. The advent of new high-throughput technology and instrumentation has now minimized the processing speed of biological samples for more complex set of conditions and produced huge information in digital data form. The amount of data generated is difficult to process, analyse, and conceptualise manually in order to draw a meaningful conclusion, so we need an algorithm-based computational approach to perform these tasks [104]. Studying and applying these analytical methods on the basis of various interdisciplinary fields including computer science, information technology, data management science, mathematics, statistics, and chemistry, working together to solve biological problems provides a broad concept of bioinformatics [48].

The exponential expansion of omics methods, which are being used in a range of different species has significantly increased molecular data collection from organisms under their native growth conditions [39, 142]. Apparently, the use of next-generation sequencing (NGS) has introduced a new age of omics approaches that further revolutionise information generation for agricultural improvement. These frameworks allow us to recognise various aspects of biomolecular organisation (genome, transcriptome, proteome, and metabolome) of complex biological systems. Obviously, these technical advances offered considerable flexibility and greater precision in the overall experimental processing and thus encouraged further research into the next generation crop improvement programmes.

Advancement in Sequencing Approaches Accelerated Bioinformatics

First-generation sequencing primarily involves the nucleotide sequencing technique suggested by Sanger and Coulson in 1975. This was based on the incorporation of selective chain-terminating dideoxynucleotides during DNA replication mediated by DNA polymerase [117]. Subsequently, Maxam and Gilbert developed a new sequencing approach based on site-specific chemical modification in nucleotide bases and subsequent cleavage adjacent to the modified base [92]. Next-generation sequencing (NGS) includes all advanced sequencing techniques that came after first-generation sequencing methods and were initiated in the year 2000 [50, 85, 97]. Next-generation sequencing has wide application including, identification of agronomically valuable trait related genes or QTLs, molecular cloning, precision breeding, search of stress related genes or QTLs and in comparative study during the crop domestication process. Detailed below are examples of advanced sequencing approaches that greatly revolutionized the field with their high precision, less technology, and affordable cost.

Illumina Sequencing

Illumina sequencing is the most popular technique nowadays for sequencing and more than 85% of sequencing is performed through this method [86]. It is also called 'Synthesis by Sequencing' (SBS) as its working principle is based on reversible fluorescent dye-terminators allowing to classify individual different bases according to their location in the sequence [89, 90]. The workflow for the Illumina sequencing comprises three major phases. (a) Preparation of a library in which nucleotide sequences are randomly fragmented into smaller sections followed by ligations of the end adapters. These short fragments are then used for PCR amplification and purification. (b) The step of cluster generation involves isothermally amplifying each fragment in the first stage of the preparation of the library. Amplified libraries are then loaded into the flow cell to complement the short stretch of DNA adapters with high density areas of immobilized oligos and a cluster created by a bridge amplification method. (c) The sequencing step is performed using the process of sequencing-by-synthesis. Specific algorithms decode each cluster's fluorescence signals to output the raw base with quality score. Illumina sequencing generates high-quality base calls that are usually acceptable above the quality threshold level [86]. The new

HiSeq and NovaSeq sequencing platforms have significantly reduced sequencing costs per base and have increased precision. Besides the Illumina platform, the Thermo Fisher Scientific Ion Proton, Ion Personal Genome Machine (PGM), Ion S5 and Ion S5 XL System, and the 454 Roche pyrosequencing platforms were also contributed significantly in the evolution of sequencing process.

Nanopore DNA Sequencing

It is the fourth-generation technology used to sequence comparatively longer nucleotide fragments in real time without the sample being amplified by PCR or chemical labeling [156]. The functioning principle is based on the electric signal modification, when nucleic acids pass through a protein-backed nanopore. The signals are decoded in real time in the form of a DNA/RNA sequence. The adapter molecule linked with DNA or RNA binds to the nanopore opening madeup of protein, and guides the nucleotide molecules to pass through it [7, 22]. This system senses changes in electrical potential as the nucleotides passes through the pore, and tracks the sequence in real time with a rate of more than 400 base per second [7, 156]. This advanced technology allows researchers to solve complex structural variants, transposons, insertions of transgenes, and analysis of repetitive regions [22]. Nanopore sequencing simplifies the assembly and annotation of the de novo genome and improves the sequenced reference genome. Moreover, epigenetic base modification is easily detected parallel to nucleotide sequencing without further protocol steps or expenses. The SmidgION, MinIÓN, GridION and PromethION platforms are now available under the Oxford Nanopore technologies [81]. The availability of powerful, lightweight and handheld sequencing instruments under this class makes it more promising and unique.

Omic Approach for Crop Improvement

The Omic tool box has opened new doors to crop plant research through a focus on the development of high yield and high stress tolerance cultivars [8]. Omics methods have facilitated the monitoring of the molecular signature changes of the living system during its growth and development under various stress conditions, including pathological, physiological, and environmental conditions [39]. These modifications are important in order to correspond with different stages of growth and important features of plants such as the ability of regeneration, flowering, fruiting, yield, senescence, and biotic/abiotic stress resistance [114, 123, 124]. The majority of biological data is produced by omic approaches to nucleic acid sequence based on recent NGS technologies [39]. The incorporation of this data into proteomics and metabolomics considerably increases the understanding about functionality of complex biological processes that are directly related to crop yield. Different omic technologies have been developed in recent years and few most significant of them are briefly discussed below [39].

Genomics

"Genomics" is typically a catalog of all genes in a given genome and involves identification, classification, and genetic sequence annotations in a particular species [20]. The availability of genomic data is now growing due to the advent of advanced sequencing technology [34]. *Arabidopsis* was the first sequenced genome in plant, later on several crops were sequenced including rice ([165]), *Amaranthus hypochondriacus* [28], *Beta vulgaris* [36], *Brassica rapa* [155], and *Glycine max* [121]. The list of agricultural plants with genome sequencing data available has been summarized in Table 1. Functional genomics approach is extremely useful for identifying and evaluating candidate genes/QTLs involved in regulation of beneficial traits including crop yield and stress tolerance [12, 139]. A new crop variety with high yield and stress tolerance could easily be produced by precision breeding based on data obtained from functional genomics [24, 151]. Apparently, genomic techniques such as large-scale parallel gene expression analysis; analysis of expressed sequence tags (ESTs), random or unique mutagenesis and mutant complementation analysis have been shown to be important in crop improvement [130, 144, 145]. Likewise, genomic databases store cross-referenced significant information obtained from various genome studies. Several databases related to plant genomic information have been developed over time [79, 167], few important of them are listed in Table 2.

Transcriptomic

Cells of an organism are genetically identical on the genomic scale but the physiological and functional properties are different because of the native regulation of the individual cell's gene expression pattern. Capturing the highly complex spatiotemporal expressions of living systems under different conditions is called transcriptomics [87, 131]. Evidently, such expression patterns of plants are environment sensitive. The high throughput RNA sequencing and easily accessible online resource with data generated from multiple studies based on environmental conditions and plant phenotype correlation is an important tool for identifying key molecular factors that regulate transcription patterns [3, 47, 64]. Transcriptome profiling may be used in identifying key regulatory genes associated with the phenotypical and physiological actions of the plant in response to certain external conditions such as biotic or abiotic stress [3, 64]. Several studies have provided valuable information on the differential expression of genes under varying

Table 1 List of agronomically important plant with available sequenced genome

Organism	Genome size	No. of genes predicted	Year	References
Brassica rapa	485 Mbp	41,174	2011	Wang et al. [155]
Beta vulgaris	714–758 Mbp	27,421	2013	Dohm et al. [36]
Chenopodium quinoa	1.39–1.50 Gb	44,776	2017	Jarvis et al. [65]
Amaranthus hypochondriacus	403.9 Mb	23,847	2016	Clouse et al. [28]
Simmondsia chinensis	887 Mb	23,490	2020	Sturtevant et al. [133]
Saccharina japonica	543.4 Mb	–	2015	Ye et al. [164]
Carica papaya	372 Mbp	28,629	2008	Ming et al. [95]
Citrullus lanatus	425 Mbp	23,440	2013	Xu et al. [162]
Cucumis melo	450 Mbp	27,427	2012	Garcia-Mas et al. [46]
Cucumis sativus	350 Mbp	26,682	2009	Huang et al. [60]
Ricinus communis	320 Mbp	31,237	2010	Chan et al. [25]
Glycine max	1115 Mbp	46,430	2010	Schmutz et al. [121]
Linum usitatissimum	~350 Mbp	43,384	2012	Wang et al. [157]
Lablab purpureus	–	20,946	2018	Chang et al. [26]
Eucalyptus grandis	691.43 Mb	–	2011	Myburg et al. [101]
Eriobotrya japonica	760.1 Mb	45,743	2020	Jiang et al. [66]
Fragaria vesca	240 Mbp	34,809	2011	Shulaev et al. [125]
Malus domestica	~742.3 Mbp	57,386	2010	Velasco et al. [148]
Prunus persica	265 Mbp	27,852	2013	Verde et al. [149]
Rubus occidentalis	43 Mbp	–	2018	VanBuren et al. [143]
Dimocarpus longan	471.88 Mb	–	2017	Lin et al. [84]
Xanthoceras sorbifolium	504.2 Mb	24,672	2019	Liang et al. [83]
Aquilaria sinensis	726.5 Mb	29,203	2020	Ding et al. [35]
Helianthus annuus	3.6 Gb	52,232	2017	Badouin et al. [14]
Lactuca sativa	2.5 Gb	38,919	2017	Reyes-Chin-Wo et al. [110]
Mentha x piperita	353 Mb	35,597	2017	Vining et al. [152]
Solanum lycopersicum	900 Mbp	34,727	2012	Sato et al. [118]
Solanum aethiopicum	1.02 Gbp	34,906	2019	Song et al. [127]
Solanum tuberosum	726 Mbp	39,031	2011	Xu et al. [161]
Capsicum annuum	~3.48 Gbp	35,336	2014	Qin et al. [108]
Oryza sativa	430 Mb	–	2002	Goff et al. [49]
Sorghum bicolor	730 Mb	34,496	2009	Paterson et al. [106]
Triticum aestivum	14.5 Gb	107,891	2018	Appels et al. [8]
Zea mays	2.3 Gb	39,656	2009	Schnable et al. [122]
Ananas comosus	382 Mb	27,024	2015	Ming et al. [96]
Musa acuminata	523 Mbp	36,542	2012	D'hont et al. [30]
Cocos nucifera	419.67 Gb		2017	Xiao et al. [160]
Phoenix dactylifera	658 Mbp	28,800	2011	Al-Dous et al. [5]

Table 2 Plant related "omics" databases

Genomic database	Brief description	References	Web address
PlantGDB	Tools and resources for Plant Comparative Genomics	Dong [37]	http://plantgdb.org/
TAIR	The *Arabidopsis* Information Resource (TAIR) is a database of genetic and molecular biology data for *Arabidopsis thaliana*	Rhee et al. [111]	https://www.arabidopsis.org/
Gramene	Curated, open-source, integrated data resource for comparative functional genomics in crops and model plant species	Gupta et al. [52]	http://gramene.org/
Plant Genome DataBase Japan (PGDBj)	Portal website dedicated to integration of plant databases based on genome information	Asamizu et al. [11]	http://pgdbj.jp/index.html?ln=en
PLAZA	Portal website and tool for comparative genomics, sequence data and an online platform to perform evolutionary analyses and data mining of green plants	Van Bel et al. [16]	https://bioinformatics.psb.ugent.be/plaza/versions/plaza_v4_dicots/
LIS—Legume Information System	Genome database dedicated for legume crop improvement	Dash et al. [31]	https://legumeinfo.org/
EnsemblPlants	Database and tool used for the annotation, analysis and display of plant genomes	Bolser et al. [19]	http://plants.ensembl.org/index.html
PGSB (Plant Genome and Systems Biology)	Database and information resource for individual plant species and provides a platform for integrative and comparative plant genome research	Spannagl et al. [129]	http://pgsb.helmholtz-muenchen.de/plant/
AgBase	It is curated resources for functional genomics and gene ontology annotation of agricultural plants and animal species	McCarthy et al. [93]	https://agbase.arizona.edu/
KEGG: Kyoto Encyclopedia of Genes and Genomes	KEGG is a database resource for understanding high-level functions and utilities of the biological system, from molecular-level information, generated by genome sequencing and other high-throughput experimental technologies	Kanehisa [68]	https://www.genome.jp/kegg/
Transcriptomic database	**Brief description**	**References**	**Web address**
TENOR (Transcriptome Encyclopedia Of Rice)	Database dedicated for RNA-Seq data of rice for different environmental stresses and plant hormone treated conditions, collects data for	Kawahara et al. [71]	https://tenor.dna.affrc.go.jp/

(continued)

Table 2 (continued)

Transcriptomic database	Brief description	References	Web address
	expression profiles, co-expression data and information of cis-regulatory elements in promoter regions		
PlantExpress	An integrated database of OryzaExpress and ArthaExpress for Gene expression networks analyses with microarray-based transcriptome data	Kudo et al. [76]	http://plantomics.mind.meiji.ac.jp/PlantExpress/
MOROKOSHI	*Sorghum bicolor* transcriptome database	Makita et al. [88]	http://matsui-lab.riken.jp/morokoshi/Home.html
CTDB	Chickpea Transcriptome Database: An integrated chickpea transcriptome database for functional and applied genomics annotation, conserved domain(s), transcription factor families, molecular markers (microsatellites and single nucleotide polymorphisms)	Verma et al. [150]	http://www.nipgr.ac.in/ctdb.html
TodoFirGene	Transcriptome database of coniferous species *Abies sachalinensis* also called Todo-fir providing useful information on functional annotations of gene	Ueno et al. [140]	http://plantomics.mind.meiji.ac.jp/todomatsu/
Proteomic database	**Brief** description	**References**	**Web address**
2P2Idb	Structural database of protein–protein complexes and their inhibitors	Basse et al. [15]	http://2p2idb.cnrs-mrs.fr/2p2idb.html
PPDB	This is a Plant Proteome database based on experimental proteome data and mass spectrometry analysis. Stores curated information about protein function, protein properties, and subcellular localization	Sun et al. [136]	http://ppdb.tc.cornell.edu/
2P2IINSPECTOR	Protein–protein interface analysis tool; efficiently characterize protein–protein and protein–ligand interfaces	Basse et al. [15]	http://2p2idb.cnrs-mrs.fr/2p2i_inspector.html
PlantPReS	Open online proteomic database for plant proteome response to stress; enables application of omics advancement in crop improvement	Mousavi et al. [100]	http://proteome.ir/
RCSB-PDB	Protein Data Bank provide resources for high-quality	Rose et al. [113]	https://www.rcsb.org/

(continued)

Table 2 (continued)

Transcriptomic database	Brief description	References	Web address
	experimental 3D structural data for proteins and complexes		
P3DB	It is a plant protein phosphorylation database	Gao et al. [45]	http://www.p3db.org/
InterPro	It is a web-based database portal that provides functional analysis of protein sequences by classifying them into families and predicting the presence of domains and important sites	Hunter et al. [63]	https://www.ebi.ac.uk/interpro/
Pfam	Database of protein families based on sequence alignments and hidden Markov models (HMMs)	Finn et al. [43]	https://pfam.xfam.org/
PhytAMP	PhytAMP is a database dedicated to antimicrobial plant peptides. It contains valuable information including taxonomic, microbiological and physicochemical data	Hammami et al. [55]	http://phytamp.pfba-lab-tun.org/main.php
PRINTS	Database of protein fingerprints i.e. group of conserved motifs	Attwood [13]	http://130.88.97.239/PRINTS/index.php
UniProtKB/Swiss-Prot	It is a high-quality annotated and non-redundant protein sequence database, based on experimental results, computed features, and scientific conclusions	Boutet et al. [21]	https://www.uniprot.org/
PDBe	Protein DataBank in Europe: Provide integrated resources of high-quality macromolecular structures deposition, annotation, and access	Velankar et al. [147]	https://www.ebi.ac.uk/pdbe/
PDBj	Protein Databank in Japan: Maintaining a structural data archive and resources description framework format, analysis tools for large structures	Kinjo et al. [73]	https://pdbj.org/
PROSITE	Protein domain database for functional characterization and annotation	Sigrist et al. [126]	https://prosite.expasy.org/
Metabolomics database	**Brief** description	**References**	**Web address**
KOMICS	The Kazusa Metabolomics Portal is a web portal based database for preprocessing, mining, and dissemination of metabolomics data	Sakurai et al. [116]	http://www.kazusa.or.jp/komics/en/
Metabolonote	Metabolonote is a database/management system that manages "metadata" for experimental data obtained through the metabolomics studies	Ara et al. [9]	http://metabolonote.kazusa.or.jp/Main_Page

(continued)

Table 2 (continued)

Transcriptomic database	Brief description	References	Web address
GMD: Golm Metabolome Database	Web-based metabolome database provides a platform to access custom mass spectral libraries, metabolite profiling experiments and related tools	Kopka et al. [74]	http://gmd.mpimp-golm.mpg.de/
PRIMe	Platform for RIKEN Metabolomics is a web-based service for metabolomics and transcriptomics. It measures standard metabolites based on NMR spectroscopy, GC/MS, LC/MS, and CE/MS	Akiyama et al. [4]	http://prime.psc.riken.jp/
PMM	RIKEN Plant Metabolome MetaDatabase (PMM) is an integrated plant metabolome data repository based on the semantic web	Sakurai et al. [115]	http://metabobank.riken.jp/pmm/db/plantMetabolomics
BiGG Models	BiGG Models contain high quality, manually-curated genome-scale metabolic models and provide a platform for browse, search, and visualize models	King et al. [72]	http://bigg.ucsd.edu/universal/metabolites
MetaboLights	MetaboLights is a database for Metabolomics experiments and derived information	Kale et al. [67]	https://www.ebi.ac.uk/metabolights/
MetaCyc	MetaCyc is a curated database of metabolic pathways based on experimental data	Karp [69] and Caspi et al. [23]	https://metacyc.org/
BMRB	Biological Magnetic Resonance Data Bank: a repository for data from NMR spectroscopy on proteins, peptides, nucleic acids, and other biomolecules	Ulrich et al. [141]	http://www.bmrb.wisc.edu/
METLIN	It comprehensive MS/MS database hosts high-quality MS/MS data at multiple energies and in pos/neg modes	Guijas et al. [51]	https://metlin.scripps.edu/landing_page.php?pgcontent=mainPage
MassBank	MassBank is a mass spectral database of high-quality experimental spectra of metabolites	Horai et al. [58]	http://www.massbank.jp/
MMCD	Madison Metabolomics Consortium Database: is a platform for metabolites and experimental spectra and provides resources for metabolomics research	Cui et al. [29]	http://mmcd.nmrfam.wisc.edu/

conditions. Relevant expression profiling tools, widely used to capture differential gene expression are

- RNA-Seq using NGS platform,
- Serial analysis of gene expression (SAGE),
- Digital Gene Expression (DGE) technique,
- Microarray Chip method,
- Real-time Quantitative PCR.

Out of these, RNA-Seq provides a high-throughput platform for transcriptome study based on the next-generation sequencing approaches [158]. The transcriptomics advancement has excelled and accelerated the genome's functional characterization. The transcriptome analysis based on RNA-Seq generates a vast amount of useful data that is not only relevant for its generator, but also always has a higher potential [158]. It is therefore highly advisable to safely store these raw or curated transcriptomic data in databases. NCBI hosts one of those high-throughput gene expression data publicly available repositories such as the Omnibus database. Table 2, lists repositories with valuable data from different studies performed till date in several plant species.

Proteomics

The protein is another large biological molecule, and in this form, all genetic information ends up in the cells as per the central dogma theory. Protein is a long chain of amino acids that are synthesided in the cell through RNA translation. Approximately, 1.5% of the genome is responsible for protein coding. Sequencing of protein is primarily intended for the identification of the primary protein structure. Proteins are the immediate regulatory switch of most cellular biochemical reactions and determine the phenotypic fate of the living cell [6, 39]. Identifying, quantifying, and characterizing these cellular proteins is highly recommended for the purpose of understanding the precise molecular mechanism and cell biophysiology. Analysis of a complete set of proteins produced by organisms, including proteins present in cells, tissues, and organs of the organism, is called proteomics under defined conditions. Protein extraction from biological samples under set conditions is the first step in proteomics followed by protein fractionation into peptides, scanning of these peptides over high-throughput MS platforms and, in the end processing the raw MS reads to obtain significant information [6]. Over the past decades, development in analytical equipment and in mass spectrometer analytic software (MS) packages has improved confidence and precision in the study of complex cellular protein mixtures in time- and cost-effective manner. The Orbitrap, Time-of-Flight (TOF) mass analyzers, triple quadrupole mass spectrometer, and Orbitrap Fusion Lumos Tribrid mass spectrometer provide significant improvement in performance and enable high-throughput proteomic analysis. In order to get relevant information to resolve biological questions, the functional annotation of large data sets obtained from proteomics is required. Bioinformatics toolkits and existing databases provide a knowledge base to gather all such relevant information from a wide range of proteomics analysis datasets and to assist in the generation and interpretation of biological processes under a profound hypothesis [6, 39, 42, 163]. Proteomic studies provide reliable ideas for the expression in native or altered conditions of a specific gene or gene collection. Moreover, it also offers information on post-translation modifications which are very important to the functionality of proteins. Additionally, it allows differential protein contents to be detected by reference in defined sets of experimental conditions. The quantitative power of proteomics analysis easily distinguishes the abundance of protein between biological systems under various physiological conditions and compares up and down expression with a set of considered conditions. Such knowledge is of great use in identifying the role of proteins in crop plants during biotic and abiotic stress [39]. Due to the ease and cost-effectiveness of the proteomics analysis, significant amounts of proteomic data from biological samples treated under different conditions have been produced in recent years. The major challenges are the annotation and archiving of these raw and processed data so that they are accessible to other researchers. Many useful tools and databases were developed to annotate and store these raw data. Analysis of functional annotation proteomics such as the Cytoscope suite with BiNGO and ClueGO, MetaCore (MetaCore TM, http:/thomsonreuters.com/metacore/), and the Integrated Interactome System (IIS) facilitate the development of biological pathways and an integrated biological network based on this dataset. Ensemble and Uniprot are important databases that are used in proteomic research. Few important proteome related databases are listed in Table 2.

Metabolomics

Analysis of the metabolic products of the living system or part of it at a given time point in a particular set of conditions is called metabolomics [53, 135]. NMR spectroscopy and mass spectrometry are common techniques for the study of these metabolites [18, 32]. Advances in these technologies provide a high-performance platform to quickly and cost-effectively analyze metabolic profiles. In general, the center of metabolomic research is the detection, quantification, and characterization of exo/endogenous metabolites in biological systems with molecular weight less than 1 kDa [54, 135]. Metabolites are immediately downstream proteome and transcriptome products and are regarded as an exact representation of phenotypic life processes, behaviors, including biotic and abiotic stress factors, under environmental conditions. Metabolomics provides a comprehensive scan of low-molecular weight molecules and metabolites in biological samples [53, 54]. High-performance analytical platforms are constantly being developed for the sorting of molecules with specific characteristics, including polarity, solubility, structure, and the presence of functional metabolic groups [135]. This new approach is optimized for filling long-awaited gapes in the system-wide study of environmental conditions and their accurate effects on plant conditions because of its ability to profile a large number of metabolites together [54]. Metabolic profiling of crops grown under a range of different conditions offers useful knowledge relating to the particular phenotypic actions of the plants and could be used to improve crops. Cellular metabolites reflect the condition upstream to the stress imposed on the living system and serve as reporter for stressors [53]. Information produced by a metabolomic scan of these reporters in a cell could be used as molecular markers for specific diseases and indicators of plant health, and can therefore be used to improve crops [53, 54]. Furthermore, if we are aware of the pathogenic nature of plants and their impact on specific metabolic pathways, we can easily scan metabolites and get ideas on disease infestation, and take advance measures to protect them. Although considerable progress has been made in the field, further studies are required to identify metabolite markers in environmental conditions for disease and other phenotypic behaviors [57, 77]. A significant source of raw and processed data for plant metabolites are metabolomic databases and few important of them are listed in Table 2.

Agri/Bioinformatics Shaping Future Agriculture

Agri-bioinformatics helps to elucidate the dynamics of genes and their networking that is useful for overall growth and developmental aspects and eventually required to improve crop or livestock [40]. Genomics contributes to agriculture by recognising and modifying genes with specific phenotypic characteristics and selecting variants accompained by markers [61]. Agri-Genomics aims to discover novel approaches by studying the genomes of crops or livestocks, gaining defensive knowledge and sustainable productivity for the food processing industry, as well as generating renewable energy and other value added goods [159]. Microbial communities associated with plants, soils, and cattle also play a major role in agriculture, as they determine the fitness of plants [56, 138] and biogeochemical properties of soil, both are responsible for yield and quality of crop [1, 103]. Due to scientific limitations and inability to grow many microbial soil species, proper analysis of such a microbiota is hampered and therefore culture-free methods help to elucidate the interaction between such microbiota and agricultural components [94, 109]. Recent studies have used metagenomics to examine the microbial diversity in the rhizosphere and the impact of external environment on the microbiome [17, 105, 128]. The role of soil bacteria in plant nutrition [107] or in the cycling of the elements can also be deciphered by the metagenomic studies [132]. Additional research could provide new knowledge about genetic and bioproduct diversity in microbes, which could be useful in understanding stress response and quorum sensing under certain environmental conditions [138, 142, 146]. Agri-bioinformatics has predictive power for innovative methods in the fields of diagnostics, tracking, and traceability in order to enhance human benefits at a lower cost and thus promote agricultural sustainability by increased knowledge about molecules and mechanisms relevant to improved agronomic traits and responses to biotic or abiotic stress management [40, 123, 124]. Increasing numbers of reference genomes in combination with lower baseline sequencing costs have also allowed genome variations to be analyzed from the discovery of single nucleotide polymorphisms (SNPs) and are able to recognize changes between different genomes may be useful to compare and classify important phenotypic characteristics and can be further used for crop improvement [123, 124]. Significant amount of polymorphism has been already reported in different crop plants and livestocks such as *A. thaliana* [44], rice [134, 161], soybean [80], tomato [2], maize [62, 78], and in animals [112] might be useful in precision breeding. Additionally, bioinformatics really helps in the proper management, application and interpretation of data during the breeding program enhances sustainable farming in a climate-changing world and helps to meet the growing demand of quality food [123, 124]. The primary goal of plant and crop biologists has been to understand and unravel the entire genetic and molecular basis of various biological processes in plants. Such understanding helps to use biological tools at lower cost to develop new cultivars with better agronomic quality [119]. The possibility of continued food supply for longer periods of time has been seriously affected by climate change and the increasing human population. The development of improved new crop varieties is therefore essential, and adaptation of existing crops to the changing climate is equally important for the continued supply of food [123, 124]. Progress in agribioinformatics has enhanced the acquisition of information, leading to the logical annotation of genes, protein and phenotypes in turn, and omics data can now be used as an important resource for the crop improvement program [39, 120] (Fig. 2). Bioinformatics and genomic tools together can accelerate the development of newly improved cultivars which might provide protection against future climate change [8, 153]. A variety of new gene-finding techniques have been applied to enhance the quality and quantity of crop plants [39, 120]. Furthermore, in the field of crop improvement, the advent of work in sequence analysis and genome annotation has played an important role [123, 124]. The increasing number of available whole genome sequences of agricultural species offers insights into their organizational genetic make-up and essential information that could be used in the crop improvement program [38]. As a consequence of the substantial advancement in the omic approaches adopted during recent years, a large amount of plant genomic data is available and can be used to functionally evaluate different genes in crop plant species [98]. In addition, the comparative omic analysis integrated with genetic engineering tools has greater potential for the development of crop varieties immune to abiotic and biotic stress (Fig. 2).

Evidently, breeding is one of the most applicable approaches since a long time in agriculture to generate new cultivars. Bioinformatics has now widely engaged in plant breeding, by using array of omics tools, plant breeding systems have been made more efficient and reliable, reducing the time and expense for developing stress-tolerant crop plants [39]. Availability of whole genome sequences due to ease of sequencing and annotation process the crop plants for which no such data are available has increased the concept of genomics based precision breeding [59, 70, 91]. Both genes and genetic variants which lead to specific agronomic characteristics

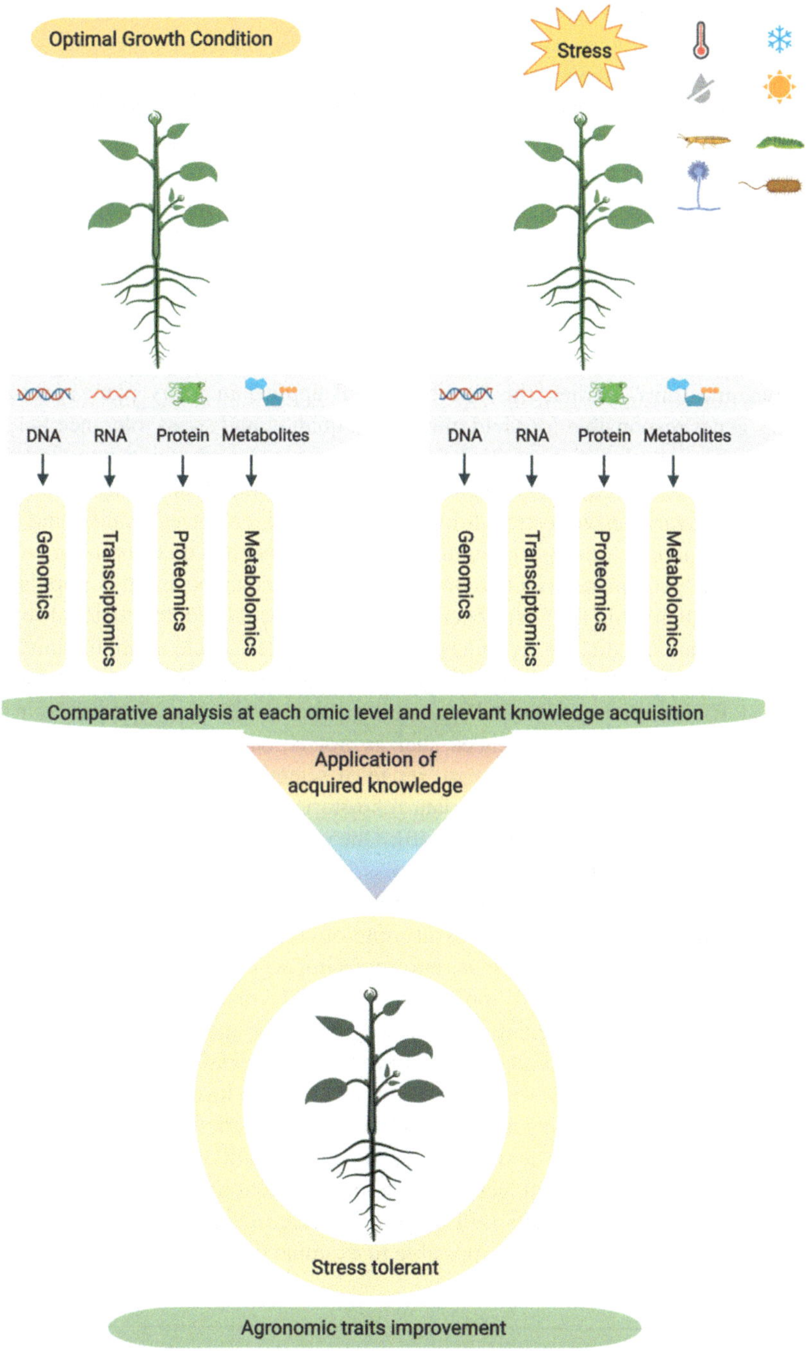

Fig. 2 The model represents the workflow for the application of Agri/Bioinformatics in crop plant improvement. This workflow involves plant growth under optimal and abiotic/biotic stress conditions (including high temperature, drought, frost, fungal, bacterial, pest infestation). Then, extraction and purification of biological samples including DNA RNA, protein and metabolites for omics

can be identified where genome sequences are available and such genotypic variations linked to phenotype can be evaluated during genomic-based breeding [59]. In every aspect of crop reproduction, as well as in quantitative trait loci (QTL) and genome-wide association studies (GWASs), the availability of genomic data to breeders is becoming increasingly relevant now. Advances in genomics have also helped to identify genetic variations in plants used in the development of climatically resilient crop varieties [99, 124]. With a wider range of phenotypes and high-resolution QTL maps, GWAS is now a favorable tool for the analysis of allelic variations in the crop plants [40]. Moreover, GWAS is also a solution to the disadvantages of existing traditional crop breeding methods [102]. GWAS has a broad range of plant breeding applications for the detection of phenotypic variability in distinctive feature and allele differences in candidate genes that tackle quantitative and dynamic traits. Earlier, this has been well applied in many plant species to identify genes responsible for yield, nutritional quality, and stress tolerance [41, 75, 82, 166]. The combination of bioinformatics with comprehensive knowledge of conventional crop breeding would undoubtedly shape farming of the next generation by generating varieties with stress tolerance, high yield, enhanced quality, and other beneficial agronomic traits [40]. Numbers of newly generated databases can be used to analyze the genome sequence, function, and genetic map position. A popular use of bioinformatics tools for next-generation farming includes computational modeling to generate predictive hypotheses using complex allele combinations and marker-based population screening to generate new variety with desired phenotype [33, 154]. Bioinformatics tools and high-performance sequencing platform have also transformed other biological sciences, such as biotechnology and molecular biology. The sequencing system spans a broad range of scientific knowledge in conjuction with a growing array of biological data repositories that have a great potential to improve traditional farming practices [40]. One of the key applications of bioinformatics in agriculture is the ability to access high resolution physical and genetic maps of the crop plants. In association with new technologies and methods, future plant breeding will achieve the necessary rate of crop improvement to ensure food security [123]. The plant breeding has enormous impact on the current crop improvement program; the successful application of bioinformatics tools will make the breeding strategy more efficient and reliable so that the growing demand for food and feed could be tackled under diversity of environmental stress. Bioinformatics toolbox, along with genome, proteome, and metabolome-level knowledge could be implemented to adapt existing and/or develop new crop varieties for completely new climatic conditions without compromising yield and quality [137]. Evidently, comparative analysis of different genomes to know functional aspects of the genes responsible for particular traits will help in future crop improvement [40]. Using bioinformatics methods, it is now possible to examine the entire genome of the crop

Fig. 2 (continued) analysis from both conditions performed. Thereafter, valuable data collected from the comparative study of these plants grown under differential conditions. Finally, the knowledge gained from this omics study could be utilized to develop new stress tolerant plant variants

plants and their association with phenotype and thus provide opportunities to modulate these crop plants with the help of advanced biotechnological and genetic engineering tools [124]. Overall, in conjunction with the implementation of biotechnology and molecular biology, it can be concluded that agribioinformatics can help in developing plant varieties with improved stress tolerance and higher yield.

Acknowledgment We are thankful to the University of California, San Diego for providing space and other facilities during the writing and submission of this book chapter.

References

1. Acosta-Martínez V, et al. Predominant bacterial and fungal assemblages in agricultural soils during a record drought/heat wave and linkages to enzyme activities of biogeochemical cycling. Appl Soil Ecol. 2014;84:69–82. https://doi.org/10.1016/j.apsoil.2014.06.005.
2. Aflitos S, et al. Exploring genetic variation in the tomato (Solanum section Lycopersicon) clade by whole-genome sequencing. Plant J. 2014;80(1):136–48. https://doi.org/10.1111/tpj.12616.
3. Agarwal P, et al. Expanding frontiers in plant transcriptomics in aid of functional genomics and molecular breeding. Biotechnol J. 2014;9(12):1480–92. https://doi.org/10.1002/biot.201400063.
4. Akiyama K, et al. PRIMe: a web site that assembles tools for metabolomics and transcriptomics. Silico Biol. 2008;8(3–4):339–45.
5. Al-Dous EK, et al. De novo genome sequencing and comparative genomics of date palm (Phoenix dactylifera). Nat Biotechnol. 2011;29(6):521–7. https://doi.org/10.1038/nbt.1860.
6. Al-Obaidi JR. Proteoinformatics and agricultural biotechnology research: applications and challenges. In: Essentials of bioinformatics, vol. III; 2019. https://doi.org/10.1007/978-3-030-19318-8_1.
7. Ansorge WJ, Katsila T, Patrinos GP. Perspectives for future DNA sequencing techniques and applications, molecular diagnostics: third edition: Elsevier Ltd; 2017. https://doi.org/10.1016/B978-0-12-802971-8.00008-0.
8. Appels R, et al. Shifting the limits in wheat research and breeding using a fully annotated reference genome. Science. 2018;361(6403) https://doi.org/10.1126/science.aar7191.
9. Ara T, et al. Metabolonote: a wiki-based database for managing hierarchical metadata of metabolome analyses. Front Bioeng Biotechnol. 2015;3(Apr):1–9. https://doi.org/10.3389/fbioe.2015.00038.
10. Arora D, et al. Use of bioinformatics in crop improvement. Biotech Today: Int J Biol Sci. 2018;8(1):88. https://doi.org/10.5958/2322-0996.2018.00001.7.
11. Asamizu E, et al. Plant genome database Japan (PGDBj): a portal website for the integration of plant genome-related databases. Plant Cell Physiol. 2014;55(1):1–7. https://doi.org/10.1093/pcp/pct189.
12. Ashikari M. Cytokinin oxidase regulates rice grain production. Science. 2005;309(5735):741–5. https://doi.org/10.1126/science.1113373.
13. Attwood TK. The PRINTS database: a resource for identification of protein families. Brief Bioinform. 2002;3(3):252–63. https://doi.org/10.1093/bib/3.3.252.
14. Badouin H, et al. The sunflower genome provides insights into oil metabolism, flowering and Asterid evolution. Nature. 2017;546(7656):148–52. https://doi.org/10.1038/nature22380.
15. Basse MJ, et al. 2P2Idb: a structural database dedicated to orthosteric modulation of protein-protein interactions. Nucleic Acids Res. 2013;41(D1):824–7. https://doi.org/10.1093/nar/gks1002.

16. Van Bel M, et al. PLAZA 4.0: an integrative resource for functional, evolutionary and comparative plant genomics. Nucleic Acids Res. 2018;46(D1):D1190–6. https://doi.org/10.1093/nar/gkx1002.
17. Bevivino A, et al. Soil bacterial community response to differences in agricultural management along with seasonal changes in a Mediterranean region. PLoS ONE. 2014;9(8) https://doi.org/10.1371/journal.pone.0105515.
18. Bhinderwala F, et al. Combining mass spectrometry and NMR improves metabolite detection and annotation. J Proteome Res. 2018;17(11):4017–22. https://doi.org/10.1021/acs.jproteome.8b00567.
19. Bolser D, et al. Ensembl plants: integrating tools for visualizing, mining, and analyzing plant genomics data. Methods Mol Biol. 2016:115–40. https://doi.org/10.1007/978-1-4939-3167-5_6.
20. Bouchez D, Höfte H. Functional genomics in plants. Plant Physiol. 1998;118(3):725–32. https://doi.org/10.1104/pp.118.3.725.
21. Boutet E, et al. UniProtKB/Swiss-Prot: the manually annotated section of the UniProt KnowledgeBase. Methods Mol Biol. 2007;406:89–112.
22. Bowden R, et al. Sequencing of human genomes with nanopore technology. Nat Commun. 2019;10(1):1–9. https://doi.org/10.1038/s41467-019-09637-5.
23. Caspi R, et al. The MetaCyc database of metabolic pathways and enzymes-a 2019 update. Nucleic Acids Res. 2020;48(D1):D455–3. https://doi.org/10.1093/nar/gkz862.
24. Cattivelli L, et al. Drought tolerance improvement in crop plants: an integrated view from breeding to genomics. Field Crops Res. 2008;105(1–2):1–14. https://doi.org/10.1016/j.fcr.2007.07.004.
25. Chan AP, et al. Draft genome sequence of the ricin-producing oilseed castor bean. Nat Biotechnol. 2010;28(9):951–6. https://doi.org/10.1038/nbt.1674.Draft.
26. Chang Y, et al. The draft genomes of five agriculturally important African orphan crops. GigaScience. 2018;8(3):1–16. https://doi.org/10.1093/gigascience/giy152.
27. Chen F, et al. The sequenced angiosperm genomes and genome databases. Front Plant Sci. 2018;9(April):1–14. https://doi.org/10.3389/fpls.2018.00418.
28. Clouse JW, et al. The Amaranth Genome: genome, transcriptome, and physical map assembly. Plant Genome. 2016;9(1) https://doi.org/10.3835/plantgenome2015.07.0062.
29. Cui Q, et al. Metabolite identification via the Madison Metabolomics Consortium Database [3]. Nat Biotechnol. 2008;26(2):162–4. https://doi.org/10.1038/nbt0208-162.
30. D'hont A, et al. The banana (Musa acuminata) genome and the evolution of monocotyledonous plants. Nature. 2012;488(7410):213–7. https://doi.org/10.1038/nature11241.
31. Dash S, et al. Legume information system (LegumeInfo.org): a key component of a set of federated data resources for the legume family. Nucleic Acids Res. 2016;44(D1):D1181–8. https://doi.org/10.1093/nar/gkv1159.
32. Deborde C, et al. Plant metabolism as studied by NMR spectroscopy. Prog Nucl Magn Reson Spectrosc. 2017;102–103:61–97. https://doi.org/10.1016/j.pnmrs.2017.05.001.
33. Dekkers JCM, Hospital F. The use of molecular genetics in the improvement of agricultural populations. Nat Rev Genet. 2002;3(1):22–32. https://doi.org/10.1038/nrg701.
34. Dhanapal AP, Govindaraj M. Unlimited thirst for genome sequencing, data interpretation, and database usage in genomic era: the road towards fast-track crop plant improvement. Genet Res Int. 2015;2015:1–15. https://doi.org/10.1155/2015/684321.
35. Ding X, et al. Genome sequence of the agarwood tree Aquilaria sinensis (Lour.) Spreng: the first chromosome-level draft genome in the Thymelaeceae family. GigaScience. 2020;9(3):1–10. https://doi.org/10.1093/gigascience/giaa013.
36. Dohm JC, et al. The genome of the recently domesticated crop plant sugar beet (Beta vulgaris). Nature. 2014;505(7484):546–9. https://doi.org/10.1038/nature12817.
37. Dong Q. PlantGDB, plant genome database and analysis tools. Nucleic Acids Res. 2004;32(90001):354D–359. https://doi.org/10.1093/nar/gkh046.

38. Ellegren H. Genome sequencing and population genomics in non-model organisms. Trends Ecol Evol. 2014;29(1):51–63. https://doi.org/10.1016/j.tree.2013.09.008.
39. Van Emon JM. The omics revolution in agricultural research. J Agric Food Chem. 2016;64 (1):36–44. https://doi.org/10.1021/acs.jafc.5b04515.
40. Esposito A, et al. Bioinformatics for agriculture in the next-generation sequencing era. Chem Biol Technol Agric. 2016;3(1):1–12. https://doi.org/10.1186/s40538-016-0054-8.
41. Famoso AN, et al. Genetic architecture of aluminum tolerance in rice (oryza sativa) determined through genome-wide association analysis and qtl mapping. PLoS Genet. 2011;7(8) https://doi.org/10.1371/journal.pgen.1002221.
42. Feist P, Hummon AB. Proteomic challenges: sample preparation techniques for Microgram-Quantity protein analysis from biological samples. Int J Mol Sci. 2015;16(2):3537–63. https://doi.org/10.3390/ijms16023537.
43. Finn RD, et al. Pfam: the protein families database. Nucleic Acids Res. 2014;42(D1):1–9. https://doi.org/10.1093/nar/gkt1223.
44. Gan X, et al. Multiple reference genomes and transcriptomes for Arabidopsis thaliana. Nature. 2011;477(7365):419–23. https://doi.org/10.1038/nature10414.
45. Gao J, et al. P3DB: a plant protein phosphorylation database. Nucleic Acids Res. 2009;37 (Suppl. 1):2008–10. https://doi.org/10.1093/nar/gkn733.
46. Garcia-Mas J, et al. The genome of melon (Cucumis melo L.). Proc Natl Acad Sci U S A. 2012;109(29):11872–7. https://doi.org/10.1073/pnas.1205415109.
47. Giacomello S, et al. Spatially resolved transcriptome profiling in model plant species. Nat Plants. 2017;3(6):17061. https://doi.org/10.1038/nplants.2017.61.
48. Gilks W. Bioinformatics: new science-new statistics? Significance. 2004;1(1):7–9. https://doi.org/10.1111/j.1740-9713.2004.00001.x.
49. Goff SA, et al. A draft sequence of the rice genome (Oryza sativa L. ssp. japonica). Science. 2002;296(5565):92–100. https://doi.org/10.1126/science.1068275.
50. Goodwin S, McPherson JD, McCombie WR. Coming of age: ten years of next-generation sequencing technologies. Nat Rev Genet. 2016;17(6):333–51. https://doi.org/10.1038/nrg.2016.49.
51. Guijas C, et al. METLIN: a technology platform for identifying knowns and unknowns. Anal Chem. 2018;90(5):3156–64. https://doi.org/10.1021/acs.analchem.7b04424.
52. Gupta P, et al. Gramene database: navigating plant comparative genomics resources. Curr Plant Biol. 2016;7–8:10–5. https://doi.org/10.1016/j.cpb.2016.12.005.
53. Hall R, et al. Plant metabolomics. Plant Cell. 2002;14(7):1437–40. https://doi.org/10.1105/tpc.140720.
54. Hall RD. Plant metabolomics: from holistic hope, to hype, to hot topic. New Phytol. 2006;169 (3):453–68. https://doi.org/10.1111/j.1469-8137.2005.01632.x.
55. Hammami R, et al. PhytAMP: a database dedicated to antimicrobial plant peptides. Nucleic Acids Res. 2009;37(Suppl. 1):963–8. https://doi.org/10.1093/nar/gkn655.
56. Haney CH, et al. Associations with rhizosphere bacteria can confer an adaptive advantage to plants. Nat Plants. 2015;1(6) https://doi.org/10.1038/nplants.2015.51.
57. Hong J, et al. Plant metabolomics: an indispensable system biology tool for plant science. Int J Mol Sci. 2016;17(6):767. https://doi.org/10.3390/ijms17060767.
58. Horai H, et al. MassBank: a public repository for sharing mass spectral data for life sciences. J Mass Spectrom. 2010;45(7):703–14. https://doi.org/10.1002/jms.1777.
59. Hu H, Scheben A, Edwards D. Advances in integrating genomics and bioinformatics in the plant breeding pipeline. Agriculture (Switzerland). 2018;8(6) https://doi.org/10.3390/agriculture8060075.
60. Huang S, et al. The genome of the cucumber, Cucumis sativus L. Nat Genet. 2009;41 (12):1275–81. https://doi.org/10.1038/ng.475.
61. Huang X, Han B. Natural variations and genome-wide association studies in crop plants. Annu Rev Plant Biol. 2014;65(1):531–51. https://doi.org/10.1146/annurev-arplant-050213-035715.

62. Hufford MB, et al. Comparative population genomics of maize domestication and improvement. Nat Genet. 2012;44(7):808–11. https://doi.org/10.1038/ng.2309.
63. Hunter S, et al. InterPro: the integrative protein signature database. Nucleic Acids Res. 2009;37(Suppl. 1):211–5. https://doi.org/10.1093/nar/gkn785.
64. Imadi SR, et al. Plant transcriptomics and responses to environmental stress: an overview. J Genet. 2015;94(3):525–37. https://doi.org/10.1007/s12041-015-0545-6.
65. Jarvis DE, et al. The genome of Chenopodium quinoa. Nature. 2017;542(7641):307–12. https://doi.org/10.1038/nature21370.
66. Jiang S, et al. Chromosome-level genome assembly and annotation of the loquat (Eriobotrya japonica) genome. GigaScience. 2020;9(3):1–9. https://doi.org/10.1093/gigascience/giaa015.
67. Kale NS, et al. MetaboLights: an open-access database repository for metabolomics data. Curr Protoc Bioinformatics. 2016;2016(March):14.13.1–14.13.18. https://doi.org/10.1002/0471250953.bi1413s53.
68. Kanehisa M. KEGG: Kyoto Encyclopedia of Genes and Genomes. Nucleic Acids Res. 2000;28(1):27–30. https://doi.org/10.1093/nar/28.1.27.
69. Karp PD. The MetaCyc Database. Nucleic Acids Res. 2002;30(1):59–61. https://doi.org/10.1093/nar/30.1.59.
70. Kaul S, et al. Analysis of the genome sequence of the flowering plant Arabidopsis thaliana. Nature. 2000;408(6814):796–815. https://doi.org/10.1038/35048692.
71. Kawahara Y, et al. TENOR: database for comprehensive mRNA-Seq experiments in rice. Plant Cell Physiol. 2016;57(1):e7. https://doi.org/10.1093/pcp/pcv179.
72. King ZA, et al. BiGG models: a platform for integrating, standardizing and sharing genome-scale models. Nucleic Acids Res. 2016;44(D1):D515–22. https://doi.org/10.1093/nar/gkv1049.
73. Kinjo AR, et al. Protein Data Bank Japan (PDBj): updated user interfaces, resource description framework, analysis tools for large structures. Nucleic Acids Res. 2017;45(D1):D282–8. https://doi.org/10.1093/nar/gkw962.
74. Kopka J, et al. GMD@CSB.DB: the Golm metabolome database. Bioinformatics. 2005;21(8):1635–8. https://doi.org/10.1093/bioinformatics/bti236.
75. Kozlov AM, Aberer AJ, Stamatakis A. ExaML version 3: a tool for phylogenomic analyses on supercomputers. Bioinformatics. 2015;31(15):2577–9. https://doi.org/10.1093/bioinformatics/btv184.
76. Kudo T, et al. Plantexpress: a database integrating OryzaExpress and ArthaExpress for single-species and cross-species gene expression network analyses with microarray-based transcriptome data. Plant Cell Physiol. 2017;58(1):e1. https://doi.org/10.1093/pcp/pcw208.
77. Kumar R, et al. Metabolomics for plant improvement: status and prospects. Front Plant Sci. 2017;8 https://doi.org/10.3389/fpls.2017.01302.
78. Lai J, et al. Genome-wide patterns of genetic variation among elite maize inbred lines. Nat Genet. 2010;42(11):1027–30. https://doi.org/10.1038/ng.684.
79. Lai K, Lorenc MT, Edwards D. Genomic databases for crop improvement. Agronomy. 2012;2(1):62–73. https://doi.org/10.3390/agronomy2010062.
80. Lam HM, et al. Resequencing of 31 wild and cultivated soybean genomes identifies patterns of genetic diversity and selection. Nat Genet. 2010;42(12):1053–9. https://doi.org/10.1038/ng.715.
81. Laver T, et al. Assessing the performance of the Oxford Nanopore Technologies MinION. Biomol Detect Quantif. 2015;3:1–8. https://doi.org/10.1016/j.bdq.2015.02.001.
82. Li H, et al. Genome-wide association study dissects the genetic architecture of oil biosynthesis in maize kernels. Nat Genet. 2013;45(1):43–50. https://doi.org/10.1038/ng.2484.
83. Liang Q, et al. The genome assembly and annotation of yellowhorn (Xanthoceras sorbifolium Bunge). GigaScience. 2019;8(6):1–15. https://doi.org/10.1093/gigascience/giz071.
84. Lin Y, et al. Genome-wide sequencing of longan (Dimocarpus longan Lour.) provides insights into molecular basis of its polyphenol-rich characteristics. GigaScience. 2017;6(5):1–14. https://doi.org/10.1093/gigascience/gix023.

85. Liu K, et al. Transcriptome analysis reveals critical genes and key pathways for early cotton fiber elongation in Ligon lintless-1 mutant. Genomics. 2012a;100(1):42–50. https://doi.org/10.1016/j.ygeno.2012.04.007.
86. Liu L, et al. Comparison of next-generation sequencing systems. J Biomed Biotechnol. 2012b;2012 https://doi.org/10.1155/2012/251364.
87. Lowe R, et al. Transcriptomics technologies. PLoS Comput Biol. 2017;13(5):e1005457. https://doi.org/10.1371/journal.pcbi.1005457.
88. Makita Y, et al. MOROKOSHI: transcriptome database in sorghum bicolor. Plant Cell Physiol. 2015;56(1):e6. https://doi.org/10.1093/pcp/pcu187.
89. Mardis ER. Next-generation sequencing platforms. Annu Rev Anal Chem. 2013;6 (1):287–303. https://doi.org/10.1146/annurev-anchem-062012-092628.
90. Maria Sirangelo T, Calabrò G. Next generation sequencing approach and impact on bioinformatics: applications in agri-food field. J Bioinform Syst Biol. 2020;03(02):32–44. https://doi.org/10.26502/jbsb.5107012
91. Matsumoto T, et al. The map-based sequence of the rice genome. Nature. 2005;436 (7052):793–800. https://doi.org/10.1038/nature03895.
92. Maxam AM, Gilbert W. A new method for sequencing DNA. Proc Natl Acad Sci U S A. 1977;74:560–4.
93. McCarthy FM, et al. AgBase: a functional genomics resource for agriculture. BMC Genomics. 2006;7:1–13. https://doi.org/10.1186/1471-2164-7-229.
94. Mendes LW, et al. Taxonomical and functional microbial community selection in soybean rhizosphere. ISME J. 2014;8(8):1577–87. https://doi.org/10.1038/ismej.2014.17.
95. Ming R, et al. The draft genome of the transgenic tropical fruit tree papaya (Carica papaya Linnaeus). Nature. 2008;452(7190):991–6. https://doi.org/10.1038/nature06856.
96. Ming R, et al. The pineapple genome and the evolution of CAM photosynthesis. Nat Genet. 2015;47(12):1435–42. https://doi.org/10.1038/ng.3435.
97. Moorthie S, Mattocks CJ, Wright CF. Review of massively parallel DNA sequencing technologies. HUGO J. 2011;5(1–4):1–12. https://doi.org/10.1007/s11568-011-9156-3.
98. Morrell PL, Buckler ES, Ross-Ibarra J. Crop genomics: advances and applications. Nat Rev Genet. 2012;13(2):85–96. https://doi.org/10.1038/nrg3097.
99. Mousavi-Derazmahalleh M, et al. Adapting legume crops to climate change using genomic approaches. Plant Cell Environ. 2019;42(1):6–19. https://doi.org/10.1111/pce.13203.
100. Mousavi SA, et al. PlantPReS: a database for plant proteome response to stress. J Proteomics. 2016;143:69–72. https://doi.org/10.1016/j.jprot.2016.03.009.
101. Myburg AA, et al. The genome of Eucalyptus grandis. Nature. 2014;510(7505):356–62. https://doi.org/10.1038/nature13308.
102. Myles S, et al. Association mapping: critical considerations shift from genotyping to experimental design. Plant Cell. 2009;21(8):2194–202. https://doi.org/10.1105/tpc.109.068437.
103. Narendra Babu A, et al. Improvement of growth, fruit weight and early blight disease protection of tomato plants by rhizosphere bacteria is correlated with their beneficial traits and induced biosynthesis of antioxidant peroxidase and polyphenol oxidase. Plant Sci. 2015;231:62–73. https://doi.org/10.1016/j.plantsci.2014.11.006.
104. Ong Q, et al. Bioinformatics approach in plant genomic research. Curr Genomics. 2016;17 (4):368–78. https://doi.org/10.2174/1389202917666160331202956.
105. Pan Y, et al. Impact of long-term N, P, K, and NPK fertilization on the composition and potential functions of the bacterial community in grassland soil. FEMS Microbiol Ecol. 2014;90(1):195–205. https://doi.org/10.1111/1574-6941.12384.
106. Paterson AH, et al. The Sorghum bicolor genome and the diversification of grasses. Nature. 2009;457(7229):551–6. https://doi.org/10.1038/nature07723.
107. Pii Y, et al. The interaction between iron nutrition, plant species and soil type shapes the rhizosphere microbiome. Plant Physiol Biochem. 2016;99:39–48. https://doi.org/10.1016/j.plaphy.2015.12.002.

108. Qin C, et al. Whole-genome sequencing of cultivated and wild peppers provides insights into Capsicum domestication and specialization. Proc Natl Acad Sci U S A. 2014;111 (14):5135–40. https://doi.org/10.1073/pnas.1400975111.
109. Rastogi G, Coaker GL, Leveau JHJ. New insights into the structure and function of phyllosphere microbiota through high-throughput molecular approaches. FEMS Microbiol Lett. 2013;348(1):1–10. https://doi.org/10.1111/1574-6968.12225.
110. Reyes-Chin-Wo S, et al. Genome assembly with in vitro proximity ligation data and whole-genome triplication in lettuce. Nat Commun. 2017;8 https://doi.org/10.1038/ncomms14953.
111. Rhee SY, et al. The Arabidopsis Information Resource (TAIR): a model organism database providing a centralized, curated gateway to Arabidopsis biology, research materials and community. Nucleic Acids Res. 2003;31(1):224–8. https://doi.org/10.1093/nar/gkg076.
112. Romiguier J, et al. Comparative population genomics in animals uncovers the determinants of genetic diversity. Nature. 2014;515(7526):261–3. https://doi.org/10.1038/nature13685.
113. Rose PW, et al. The RCSB Protein Data Bank: new resources for research and education. Nucleic Acids Res. 2013;41(D1):475–82. https://doi.org/10.1093/nar/gks1200.
114. Saad MG, et al. Algal biofuels: current status and key challenges. Energies. 2019;12 (10) https://doi.org/10.3390/en12101920.
115. Sakurai T, et al. PRIMe update: innovative content for plant metabolomics and integration of gene expression and metabolite accumulation. Plant Cell Physiol. 2013;54(2):e5. https://doi.org/10.1093/pcp/pcs184.
116. Sakurai T, et al. A single blastocyst assay optimized for detecting CRISPR/Cas9 system-induced indel mutations in mice. BMC Biotechnol. 2014;14:1–11. https://doi.org/10.1186/1472-6750-14-69.
117. Sanger F, Coulson AR. A rapid method for determining sequences in DNA by primed synthesis with DNA polymerase. J Mol Biol. 1975;94(3):441–8. https://doi.org/10.1016/0022-2836(75)90213-2.
118. Sato S, et al. The tomato genome sequence provides insights into fleshy fruit evolution. Nature. 2012;485(7400):635–41. https://doi.org/10.1038/nature11119.
119. Schlueter JA, et al. Mining EST databases to resolve evolutionary events in major crop species. Genome. 2004;47(5):868–76. https://doi.org/10.1139/G04-047.
120. Schlueter SD, Dong Q, Brendel V. GeneSeqer@PlantGDB: gene structure prediction in plant genomes. Nucleic Acids Res. 2003;31(13):3597–600. https://doi.org/10.1093/nar/gkg533.
121. Schmutz J, et al. Genome sequence of the palaeopolyploid soybean. Nature. 2010;463 (7278):178–83. https://doi.org/10.1038/nature08670.
122. Schnable PS, et al. The B73 maize genome: complexity, diversity, and dynamics. Science. 2009;326(5956):1112–5. https://doi.org/10.1126/science.1178534.
123. Shafi A, et al. Impact of bioinformatics on plant science research and crop improvement. In: Essentials of bioinformatics, vol. III; 2019. p. 29–46. https://doi.org/10.1007/978-3-030-19318-8_2.
124. Shafi A, Zahoor I. Bioinformatics and plant stress management. In: Essentials of bioinformatics, vol. III; 2019. p. 47–78. https://doi.org/10.1007/978-3-030-19318-8_3.
125. Shulaev V, et al. The genome of woodland strawberry (Fragaria vesca). Nat Genet. 2011;43 (2):109–16. https://doi.org/10.1038/ng.740.
126. Sigrist CJA, et al. PROSITE, a protein domain database for functional characterization and annotation. Nucleic Acids Res. 2009;38(Suppl. 1):161–6. https://doi.org/10.1093/nar/gkp885.
127. Song B, et al. Draft genome sequence of Solanum aethiopicum provides insights into disease resistance, drought tolerance, and the evolution of the genome. GigaScience. 2019;8(10):1–16. https://doi.org/10.1093/gigascience/giz115.
128. Souza RC, et al. Shifts in taxonomic and functional microbial diversity with agriculture: how fragile is the Brazilian Cerrado? BMC Microbiol. 2016;16(1):1–15. https://doi.org/10.1186/s12866-016-0657-z.

129. Spannagl M, et al. PGSB plantsDB: updates to the database framework for comparative plant genome research. Nucleic Acids Res. 2016;44(D1):D1141–7. https://doi.org/10.1093/nar/gkv1130.
130. Sreenivasulu N, Sopory SK, Kavi Kishor PB. Deciphering the regulatory mechanisms of abiotic stress tolerance in plants by genomic approaches. Gene. 2007;388(1–2):1–13. https://doi.org/10.1016/j.gene.2006.10.009.
131. Stahl PL, et al. Visualization and analysis of gene expression in tissue sections by spatial transcriptomics. Science. 2016;353(6294):78–82. https://doi.org/10.1126/science.aaf2403.
132. Stempfhuber B, et al. Spatial interaction of archaeal ammonia-oxidizers and nitrite-oxidizing bacteria in an unfertilized grassland soil. Front Microbiol. 2016;6(Jan):1–15. https://doi.org/10.3389/fmicb.2015.01567.
133. Sturtevant D, et al. The genome of jojoba (Simmondsia chinensis): a taxonomically isolated species that directs wax ester accumulation in its seeds. Sci Adv. 2020;6(11):1–14. https://doi.org/10.1126/sciadv.aay3240.
134. Subbaiyan GK, et al. Genome-wide DNA polymorphisms in elite indica rice inbreds discovered by whole-genome sequencing. Plant Biotechnol J. 2012;10(6):623–34. https://doi.org/10.1111/j.1467-7652.2011.00676.x.
135. Sumner LW, Mendes P, Dixon RA. Plant metabolomics: large-scale phytochemistry in the functional genomics era. Phytochemistry. 2003;62(6):817–36. https://doi.org/10.1016/S0031-9422(02)00708-2.
136. Sun Q, et al. PPDB, the Plant Proteomics Database at Cornell. Nucleic Acids Res. 2009;37(Suppl. 1):969–74. https://doi.org/10.1093/nar/gkn654.
137. Takeda S, Matsuoka M. Genetic approaches to crop improvement: responding to environmental and population changes. Nat Rev Genet. 2008;9(6):444–57. https://doi.org/10.1038/nrg2342.
138. Timmusk S, et al. Drought-tolerance of wheat improved by rhizosphere bacteria from harsh environments: enhanced biomass production and reduced emissions of stress volatiles. PLoS ONE. 2014;9(5) https://doi.org/10.1371/journal.pone.0096086.
139. Tuberosa R. Mapping QTLs regulating morpho-physiological traits and yield: case studies, shortcomings and perspectives in drought-stressed maize. Ann Bot. 2002;89(7):941–63. https://doi.org/10.1093/aob/mcf134.
140. Ueno S, et al. TodoFirGene: developing transcriptome resources for genetic analysis of abies sachalinensis. Plant Cell Physiol. 2018;59(6):1276–84. https://doi.org/10.1093/pcp/pcy058.
141. Ulrich EL, et al. BioMagResBank. Nucleic Acids Res. 2008;36(Suppl. 1):402–8. https://doi.org/10.1093/nar/gkm957.
142. Urano K, et al. "Omics" analyses of regulatory networks in plant abiotic stress responses. Curr Opin Plant Biol. 2010;13(2):132–8. https://doi.org/10.1016/j.pbi.2009.12.006.
143. VanBuren R, et al. A near complete, chromosome-scale assembly of the black raspberry (Rubus occidentalis) genome. GigaScience. 2018;7(8) https://doi.org/10.1093/gigascience/giy094.
144. Varshney R, Graner A, Sorrells M. Genomics-assisted breeding for crop improvement. Trends Plant Sci. 2005;10(12):621–30. https://doi.org/10.1016/j.tplants.2005.10.004.
145. Varshney RK, Hoisington DA, Tyagi AK. Advances in cereal genomics and applications in crop breeding. Trends Biotechnol. 2006;24(11):490–9. https://doi.org/10.1016/j.tibtech.2006.08.006.
146. Vayssier-Taussat M, et al. Shifting the paradigm from pathogens to pathobiome new concepts in the light of meta-omics. Front Cell Infect Microbiol. 2014;5(Mar):1–7. https://doi.org/10.3389/fcimb.2014.00029.
147. Velankar S, et al. PDBe: Protein Data Bank in Europe. Nucleic Acids Res. 2012;40(D1):445–52. https://doi.org/10.1093/nar/gkr998.
148. Velasco R, et al. The genome of the domesticated apple (Malus × domestica Borkh.). Nat Genet. 2010;42(10):833–9. https://doi.org/10.1038/ng.654.

149. Verde I, et al. The high-quality draft genome of peach (Prunus persica) identifies unique patterns of genetic diversity, domestication and genome evolution. Nat Genet. 2013;45 (5):487–94. https://doi.org/10.1038/ng.2586.
150. Verma M, et al. CTDB: an integrated chickpea transcriptome database for functional and applied genomics. PLoS ONE. 2015;10(8):1–10. https://doi.org/10.1371/journal.pone.0136880.
151. Vij S, Tyagi AK. Emerging trends in the functional genomics of the abiotic stress response in crop plants. Plant Biotechnol J. 2007;5(3):361–80. https://doi.org/10.1111/j.1467-7652.2007.00239.x.
152. Vining KJ, et al. Draft genome sequence of Mentha longifolia and development of resources for mint cultivar improvement. Mol Plant. 2017;10(2):323–39. https://doi.org/10.1016/j.molp.2016.10.018.
153. Visendi P, et al. An efficient approach to BAC based assembly of complex genomes. Plant Methods. 2016;12(1):1–9. https://doi.org/10.1186/s13007-016-0107-9.
154. Walsh B. Quantitative genetics in the age of genomics. Theor Popul Biol. 2001;59(3):175–84. https://doi.org/10.1006/tpbi.2001.1512.
155. Wang X, et al. The genome of the mesopolyploid crop species Brassica rapa. Nat Genet. 2011;43(10):1035–40. https://doi.org/10.1038/ng.919.
156. Wang Y, Yang Q, Wang Z. The evolution of nanopore sequencing. Front Genet. 2014;5 (Dec):1–20. https://doi.org/10.3389/fgene.2014.00449.
157. Wang Z, et al. The genome of flax (Linum usitatissimum) assembled de novo from short shotgun sequence reads. Plant J. 2012;72(3):461–73. https://doi.org/10.1111/j.1365-313X.2012.05093.x.
158. Wang Z, Gerstein M, Snyder M. RNA-Seq: a revolutionary tool for transcriptomics. Nat Rev Genet. 2009;10(1):57–63. https://doi.org/10.1038/nrg2484.
159. Wilson SA, Roberts SC. Metabolic engineering approaches for production of biochemicals in food and medicinal plants. Curr Opin Biotechnol. 2014;26:174–82. https://doi.org/10.1016/j.copbio.2014.01.006.
160. Xiao Y, et al. The genome draft of coconut (Cocos nucifera). GigaScience. 2017;6(11):1–11. https://doi.org/10.1093/gigascience/gix095.
161. Xu X, et al. Genome sequence and analysis of the tuber crop potato. Nature. 2011;475 (7355):189–95. https://doi.org/10.1038/nature10158.
162. Xu Y, et al. The draft genome of watermelon (Citrullus lanatus) and resequencing of 20 diverse accessions. Nat Genet. 2013;45(1):51–8. https://doi.org/10.1038/ng.2470.
163. Yates JR. Recent technical advances in proteomics. F1000Research. 2019;8:1–8. https://doi.org/10.12688/f1000research.16987.1
164. Ye N, et al. Saccharina genomes provide novel insight into kelp biology. Nat Commun. 2015;6 https://doi.org/10.1038/ncomms7986.
165. Yu J. A draft sequence of the rice genome (Oryza sativa L. ssp. indica). Science. 2002;296 (5565):79–92. https://doi.org/10.1126/science.1068037.
166. Zhang Z, et al. Improving the accuracy of whole genome prediction for complex traits using the results of genome wide association studies. PLoS ONE. 2014;9(3):1–12. https://doi.org/10.1371/journal.pone.0093017.
167. Zhang Z, et al. Database resources of the National Genomics Data Center in 2020. Nucleic Acids Res. 2019; https://doi.org/10.1093/nar/gkz913.

Digital Marketing: A Sustainable Way to Thrive in Competition of Agriculture Marketing

Subhas Chandra Bose and Ravi Kiran

Introduction

The agriculture and food sector is a vital element of any economy as it determines the existential potentiality of any community by taking care of the primary need of human being and its livestock. The share of agriculture in the total GDP of any economy is quite significant and even surpasses the half way mark in the case of developing nations. Its employment providing capability is well appreciated in economic literature as the percentage of population employed in this sector is highest in countries like India. As economy develops, there is a decline in the share of agriculture in GDP and employment generation capability, making it less attractive as an economic sector. Agriculture as an economic sector is also facing new challenges, such as biodiversity protection, broad-based cultural landscape preservation, rural development including the creation and safeguarding of jobs, and the notion of regional products as cultural assets, the impact of climate change, etc.

To remain significant, agriculture sector has undergone changes which facilitated the adoption of new business strategies which can lead to the expansion of the international market of agricultural products. The contribution of agriculture can be improved by strengthening relations in the industry, through agro processing, value additions and improving post-harvest operation, storages, distribution and logistics that are essential elements of agribusiness value chains. In an economy like USA, with highly productive agriculture sector, the agribusiness sector is large consisting of different layers of activity in marketing, storage and processing. Productivity is the key factor affecting growth in agribusiness. The growth and productivity of agribusiness are vital for economic development of any country and the new forces

S. C. Bose (✉) · R. Kiran
School of Humanities and Social Sciences, Thapar Institute of Engineering and Technology, Patiala, Punjab, India
e-mail: scbose@thapar.edu

A. K. Upadhyay et al. (eds.), *Bioinformatics for agriculture: High-throughput approaches*, https://doi.org/10.1007/978-981-33-4791-5_8

for economic growth worldwide are technological innovation in every aspect of the business processes [8].

The agriculture and food sector is facing multiple challenges like significant increase in the demand for food, the availability of natural resources such as fresh water and productive arable land. To achieve the UN Sustainable Development Goal of a 'world with zero hunger' by 2030 there is a requirement of more productive, efficient, sustainable, inclusive, transparent and resilient food systems (FAO 2017, p. 140). This requires the transformation of the current agri-food system.

One of the key strategies adopted by the policy makers in the area of agriculture is modern marketing strategy. Marketing affects everybody because, as consumers and producer, we cannot escape the market, even those of us who try to live the simple life [2–4]. Proper identification of consumer's need, new product development, creating awareness about such value added products, developing efficient distribution system and providing post sales services requires adequate marketing strategies. Use of marketing tools and techniques for improving the performance of agriculture sector is well accepted as expert believes that marketing is everything [11].

As researchers believe that marketing is marketing, irrespective of the product or marketplace so presence of marketing activities in agriculture is quite common. Yet this sector has some unique features which make it distinct from other consumer goods. The agriculture sector usually takes Business-to-Business approach, i.e. market segmentation, analysis, and instituting a marketing plan, rather than targeting individual customers [1]. Secondly, the perishable nature of agricultural products and their huge volume requires exceptional and apt transportation as delay can damage the product which will impact the customer adversely [7]. To overcome the challenges, there is a need of adopting modern and efficient technologies like digital innovation.

'Fourth Industrial Revolution' is observing numerous sectors rapidly altered by 'disruptive' digital technologies such as Blockchain, Internet of Things, Artificial Intelligence and Immerse Reality. In agriculture sector, similar transformation is observed in the form of spread of mobile technologies, remote-sensing services and distributed computing. Expert predicts that a new revolution in agriculture sector is taking place and they call it 'digital agricultural revolution'. This will be the newest shift which could help ensure agriculture meets the needs of the global population into the future.

Food and Agriculture Organization of the United Nation (2019) predicted that 'Digitalization' of agriculture sector will lead to modification in every aspect of the agri-food chain as management of resources all over the system will be optimized, customized, intelligent and pre-emptive. It will perform in real time in a hyper-connected way, driven by data. Value chains will become traceable and coordinated at the most detailed level whilst different fields, crops and animals can be accurately managed to their own optimal preparations. Digital agriculture is expected to form arrangements that are highly productive, pre-emptive and compliant to changes. This, in turn, could lead to greater food security, profitability and sustainability. The inclusive application of the Internet of Things and e-commerce and the online transaction of agricultural products have effectively improved the online marketing efficiency of agricultural products, effectively expanded the marketing channels for agricultural products, and significantly reduced the marketing costs of products [7].

Agriculture Marketing

Marketing activities deal with the identification of customer's need, creation of products to satisfy such needs and exchanging such products with potential buyers at a price thereby establishing profitable relationships between seller and buyer. This helps in promoting and enabling the process of trading goods or service. As marketing is involved in every economic sector, its presence in agriculture sector is very prominent and responsible for the changes which have taken place in this area. A sound marketing strategy ensures reasonable benefits to the producers and consumers [6]. Agriculture marketing deals with the application of marketing tools and techniques in the area of agriculture with the main objective of increasing the value of the agricultural output and thereby maximizing profit. Marketing has helped in increasing the satisfaction level of agriculture customers through proper identification of their need and delivering such benefits as desired by them. The most laudable contribution of agriculture marketing is timely decimation of accurate information to potential customers. This has facilitated both the buyer and seller to plan their action for better outcome. So agricultural marketing covers the amenities involved in transferring agricultural outputs from the farm to the ultimate users. Several interrelated actions are involved in undertaking this, such as planning, production, grading, packing, transport, storage, agro-and food processing, distribution, advertising and sale.

The marketing itself has undergone several changes in the last two decades mainly due to technological progresses. The impact of disruptive technologies in every aspect of human life is quite evident. The technological revolution, which has completely changed the structure of industrial production in recent decades, is coming to agriculture. Changes occur in the market, in the organization of agricultural production, in the structure of consumption, and in the system of agro-innovations [13]. So to remain competitive in the market and economically viable, people involved in agriculture sector are also adopting new technologies for improving their performance. So the agricultural sector needs well-functioned market to drive growth, employment and economic prosperity and in order to provide efficiency and effectiveness in the agriculture marketing system, adoption of new marketing tools and mode of transaction is required [14, 15].

Many studies in this area have listed a number of challenges associated with marketing of agricultural harvest: like farmers having inadequate market information about agricultural products; low literacy level among the farmers particularly in developing countries; multiple channels of distribution eating away the profit share of farmers. There are too many intermediaries who share the farmer's profit. So the impact of technologies in the rural area is limited and beyond the reach of poor farmers [10]. All these factors promote the application of new technologies in the field of agriculture. The main objective is to overcome the causes responsible for poor performance of this sector at the one hand and to face the challenges that the future global agriculture market is going to pose.

Digitalization of Agriculture Marketing

Digital marketing is a non-conventional virtual platform basically on Internet for identifying and accepting consumer's requirements, promoting goods and services, associating customers through usage of digital technologies and devices [9]. Nowadays running a business without any online presence is not possible as because figures show that the trend from consuming in shops goes in the direction of online consuming [12]. The agriculture sector being so vital for the economy cannot be devoid of the digitalization process.

In simple term, digital agriculture marketing is the application of digital technologies in agriculture marketing and involves promotion of agricultural products or brands via one or more forms of electronic media. The main objective of digital marketing is to promote brands, build preference and increase sales through various digital marketing techniques [1]. As indicated by Hooker [5] there is a massive increase in the number of agribusinesses which are looking for the Internet as a marketing, management, service and coordination tool.

For anyone who wants to go for digital marketing should have individual website or App through which different promotion and tools are linked and used. Digital marketing can help an organization to reach its target customer easily and at lowest possible cost. Usually, above-the-line promotion, refers to traditional methods of advertising, such as, print advertisements in magazines and newspapers, billboards and TV advertisements is very expensive and increases the cost of marketing. In contrast, below-the-line promotion aims to reach more targeted groups of consumers that too at very low cost. Use of digital technology helps in below-the-line promotion.

E-marketing is also known as Internet Marketing, Web Marketing, Digital Marketing, or Online Marketing and referred to those marketing strategies and techniques which use online ways to reach target customers, gather relevant information and provide the required offerings. In the whole process, digital technologies are used. E-marketing of agricultural products means marketing of agricultural products through online ways from agricultural producer to any business organizations or final users (Bhosage 2018). E-marketing or digital marketing of agricultural products is gaining acceptance worldwide due to its ability to overcome some of the major problems associated with agriculture, right from production till final consumption.

Enablers of Digital Marketing of Agricultural Products

FAO (2018) report on digital marketing of agricultural products has identified three key factors (enablers) which are responsible for the growth of digital marketing of agriculture and will influence the future growth. They are:

1. The increase use of Internet and mobile and social networks among farmers and agricultural extension officers,
2. Digital skills among the rural population and
3. A culture which encourages digital agri-preneurship and innovation.

High speed Internet connections, i.e. 4G connection, smartphones, mobile apps, social media and digital engagement platforms have significantly improved access to information and services for those who are involved in agriculture production and distribution. With newly acquired skill in digital technology, people find it convenient to use it. Secondly, the reach and 24X7 availability have contributed positively. Education and income are two important determinants as educated people tend to adopt new technologies faster and better. Digitalization increases the demand for digital skill development. So with the rise in literacy rate, the demand for digital skill increases which leads to faster adoption. With the rise in competence level of the potential users, the process of digitalization will accelerate.

Advantages of Digitalization

Different studies in this area have established that digitalization has the potential to deliver significant economic, social and environmental benefits. It is capable of overcoming many challenges which the tradition processes can't do. Due to its unconventional approach, it maximizes the reach among its target users, thereby delivering the result which other mode can't. The study carried out by Jiang Zhao (2019) confirms that use of digital technologies for marketing agricultural product is successful as it contributes positively in developing targeted product positioning which helps in creating product differentiation. The development of e-commerce, the new transaction mode based on technological innovation is forming a powerful force, which promotes the transformation of the entire agricultural sector as it has enhanced agricultural competitiveness and expanded the international market of agricultural products.

According to Bouris et al. there are two important aspects to the marketing of agricultural products. The first deals with the physical process that brings products from producers to consumers which includes the collection, packaging, transport, processing, storage and the retail sale of agricultural products. The second aspect involves the market pricing mechanism. Digitalization is capable of improving the performance of these stages and upon its application can benefit the agriculture sector by creating huge opportunities for farmers and other people involved in this sector. The advantages can be broadly grouped under the following:

(A) Market Expansion

Digitalization of agriculture marketing will help in widening the market. With the application of Internet, farmers can approach worldwide market with increased number of potential customers. With easy and free flow of information, awareness level is bound to increase, which will facilitate the smooth and hassle free transaction of agricultural output. With the spread of Internet in the rural areas, the availability of information regarding different aspect of agriculture and its marketing is assisting in developing this market. Internet has made this market operational 24X7, which enables the farmers, the most important stakeholders, access the market as per their convenience, which in turn increasing the size of the market.

In fact, the digitalization process is creating an environment of freedom to do business and people can participate in the transaction process as per their convenient time as every time is the real time. It is much easier for customers to find substitutes from competitors on the Internet because of the minimal effort a person has to make to get to another website (Schwarz et al. 2015). So it can be said that digitalization allows people to participate not only as per their suitable time, it also provides huge amount of information which makes the whole process more effective and efficient.

(B) Cost Reduction

The main objective of marketing is not only to facilitate transaction of goods and services but also to carry out these activities at minimum cost. Use of Internet has increased the reach of this market in the one hand and has also reduced the transaction cost. Schwarz et al. (2015) found that average return on online marketing investment is very high making the whole process quite profitable which has attracted many agribusiness owners to adopt Internet marketing. Easy access to information reduces the risks of overproduction of agricultural crops, provides access to real prices for agricultural products, and reduces the cost of intermediary services, simplifies the construction of transport chains. Shortening the chain of agricultural product marketing through digital marketing helps in saving time and avoiding unwanted expenses [7].

With the use of digital technology, customer information is relatively easy and inexpensive to gather, cheap to store, and fruitful to mine. A properly planned and effectively targeted e-marketing campaign can reach the right customers at a much lower cost than traditional marketing methods (Tsekouropoulos 2011). So companies discover in the Internet a source to reduce customer-service costs, which helps in sustaining customer relationships, to cover marketing messages personally and thus enable mass customization (Johnson 2002). Digitalization creates the required infrastructure which facilitates the reduction in marketing cost, which is crucial for increasing the income of farmers and reducing cost to the consumers.

(C) Reduction in Agriculture Waste

With the application of digital technologies, the transaction process has not only become smooth but less time taking. The volume of agricultural products is huge and is easily perishable. Time lost due to complicated transaction process actually contributes in enhancing the chances of wastage and degradation of output. Application of digital technology makes the transaction process smooth, fast and accurate, which actually reduces the wastage of agricultural products. The net income of the farmers and other people involved in agriculture marketing increases due to decrease in wastage. Establishment of online agriculture market facilitates people to sell or buy agricultural product and provides platform to advertise their outputs.

(D) Encourages Healthy Competition

Faced with hundreds of millions of consumers and tens of thousands of sellers, it is self-evident that the competition for agricultural products is very fierce. As the gap between consumers and agricultural producer's increases, the idea of building trust and understanding between these two parties is crucial (Perkins 2010). Digitalization has provided an efficient platform for gathering information about different product availability, their features, sources and price. It acts as a repository for the buyers and seller which can easily access and used. It is much easier for customers to find substitutes from competitors on the Internet because of the minimal effort a person has to make to get to another website. Every Internet page is full of different kinds of touch points as advertisements and offers which makes it hard for companies to guide potential buyers to a certain website without "losing" them through their way of the Internet. Online platform encourages customers to share their experience of product utilization against the experience of using competitors' products and then to publish these blocks of information online. So it encourages consumers to get involved in the marketing process and thereby increases the credibility of the message delivered (Riz 2013).

Digitalization encourages higher involvement of customers and producers in the marketing process of agricultural output which give rise to healthy competition. With reasonable chance for comparison, the customer can make their purchase without any bias and the producer can identify the scope for improving the quality of their output. Many small producers can join the transaction process more easily which was otherwise not possible in the traditional method. Producers can comprehend the need for improving the packaging of their product and can avail the advantage of branding their output. Companies from developing countries involved in agriculture marketing have used digital technology to target customers of developed nations and were able to sell their output at higher price. Small Asian countries like Malaysia, Thailand and Indonesia have gained a lot from this process.

(E) Easy Availability of Rare Products

The digitalization of agriculture has encouraged the producers to market their output online which have increased the availability of those products which were earlier geographically confined and rarely available in the open market like 'Dates' of Arabia, tea of Darjeeling, Kiwi of New Zeeland, etc. This have encouraged farmers of these products to increase their output and selling it online. Many companies involved in online marketing of these products have added them in their product line with proper branding.

In fact, digital technology can be used in every aspect of agriculture right from plantation to final consumption by users. It is being used by agriculturist to predict the expected pests attach in the area of cultivation and the potential need of pest control agents. The cultivators can analyse the expected yield from the standing crops and can plan mode of action for increasing the return by proper utilization of this technology. This technology can assist the marketers of agricultural products to deliver the right kind of product to the appropriate customer at the right place with minimum effort and cost. If properly applied digital technology can bring paradigm shift in the area of agriculture and food sector.

Though the potential role of digital technology in the area of agriculture and its marketing seems impressive but a lot has to be done to make it more effective. FAO (2019) report on digitalization of agriculture and its marketing reflects the positive aspect on the basis of falling handset prices (smart phones), increasing Internet coverage and the growing youth population. This indicates the significant opportunities for the use of mobile phones in agricultural areas and the huge opportunity for performance improvement. The same report also expressed factors which can prevent the achievement of the ultimate goal.

Challenges and Future Course of Action

There are several challenges involved in marketing of agricultural produce. Bojkic et al. [1] states 'there are too many vultures that eat away the benefits that the farmers are supposed to get ....' The study also highlights the presence of several loopholes in the present system and identified the absence of organized and regulated marketing system for marketing the agricultural produce.

The limited presence of Internet and smart phone in developing countries in general and rural area in particular is the major hindrance in en-cashing the real potential of digitalization. Low income and lack of awareness can be considered for such situation. On the other hand, farmers are not very quick in adopting new technologies for usage in agriculture. So any strategy to increase digitalization of agricultural sector should consider developing required infrastructure first. This will create opportunities to access information and services through mobile applications, online videos and social media.

The low literacy level in the rural area is another factor responsible for poor adoption rate. FAO (2019) in its annual report have expressed grave concern over the low level of digital skills and e-literacy and considered it as a significant constraint to the use of new technologies. To increase the digital skill and e-literacy, it is necessary to increase the literacy level among youths. School administration should take initiative to include in their curricula the digital subjects, for improving knowledge and skills among teachers/students. So for complete adoption of this technology, the concern authorities should work in creating opportunity for increasing literacy rate as well as enhancing digital skills simultaneously and continuously. Making people aware about the benefits of such technologies require boost. Education and supporting services must be improved to enhance the acceptance of digital technologies (FAO 2019).

The role of government in establishing and managing digital environment is very vital and the successful implementation of digital programs heavily depends on government initiatives. The huge investment required for this program can be procured only through active involvement of government and private sector. Establishment of required infrastructure and regulatory authorities can only be achieved successfully through active government participation. There is increased interest in data-enabled farming and related services and many new entrants from the technology industry and start-ups. Encouraging start-up to enter in this area can significantly increase the usage of new technologies which can boost the performance of this sector. In fact, agriculture marketing is in dire need of some breakthrough technology which can increase the productivity and efficiency of the agricultural sector. Many researchers and policy makers consider digital technology as the only source of such drastic change.

Conclusion

Agriculture sector is an important economic and social sector, yet the poor performance of this sector remains to be the major source of concern for the policy makers. Though there is an extensive use of technology in different aspects of agriculture sector in the last few years to increase productivity, but due to its unique characteristics the expected result is not very satisfactory. This encourages people involved in agriculture marketing to apply new technologies like digital technology. Digitalization has been successfully applied in several sectors with outstanding results like improvement in customer's satisfaction level, decrease in transaction cost, fair competition and increased information availability.

This has encouraged the agriculturist, policy makers and researcher to apply digital technology in this area. Though applied in limited areas yet the outcome is encouraging and has increased the level of interest in it. With the application of digital technology in agriculture marketing, the performance of this sector is becoming promising. Rising awareness about digitalization among potential users is making this process more attractive and indispensible. A right doze of digital marketing supported by embedding of technology can go a long way in enhancing the significance of this sector. This study has pointed a few steps to enhance the impact of digitalization marketing of agricultural products.

Companies involved in digital agriculture marketing finds the process helpful in increasing their customer reach, better availability of information, reduction in the cost of transaction, increased customer satisfaction and decreased wastage. These factors make these companies more efficient and profitable in compare to those who are yet to adopt the new technology. With the elimination of the factors responsible for creating hindrance, it is expected that more agriculture based companies can go for digitalization of their marketing activities.

References

1. Bojkic V, et al. Digital marketing in agricultural sector. ENTRENOVA 8–9, September 2016; 2016. p. 136–41.
2. Bouris J, et al. Agricultural marketing competitive strategies and innovative practices in Greece. International Scientific Conference eRA6, ISSN-1791-1133; 2006.
3. Hill RP. Consumer culture and the culture of poverty: implications for marketing theory and practice. Mark Theory. 2002;2(3):273–93.
4. Holt DB. Does cultural capital structure American consumption? J Consum Res. 1998;25(1)
5. Hooker NH, Heilig J, Ernst S. What is unique about E-agribusiness?, World Food and Agribusiness Symposium; 2001.
6. Jerome S. A study on agricultural marketing strategies and challenges faced by the Ponmalai Santhai (local market) farmers in Tiruchirappalli. Int J Econ Manag Stud. 2017;4(9):15–20.
7. Juswadi J, Sumarna P, Mulyati NS. Digital marketing strategy of Indonesian agricultural products. Adv Soc Sci Educ Humanit Res. 2019;429:105–10.
8. Lalkaka R, Abetti PA. Business incubation and Enterprise support system in restructuring countries. Creat Innov Manag. 1999;8(3):197–209.
9. Mishra CK. Digital marketing: scope opportunities and challenges. 2000. https://doi.org/10.5772/intechopen.92329.
10. Rajendran G, Karthikesan P. Agricultural marketing in India-an overview. Asia Pac J Res. 2014;I(XVII)., ISSN: 2320-5504, E-ISSN-2347-4793
11. Saren M. Marketing is everything: the view from the street. Mark Intell Plan. 2007;25(1):11–6.
12. Schwarzl S, Grabowska M. Online marketing strategies: the future is here. J Int Stud. 2015;8(2):187–96. https://doi.org/10.14254/2071-8330.2015/8-2/16.
13. Sulimin VV, et al. Digitization of agriculture: innovative technologies and development models. IOP Conf Ser: Earth Environ Sci. 2019;341:012215.
14. Trendov NM, Varas S, Zeng M. Digital technologies in agriculture and rural areas. Rome: Food and Agriculture Organization of the United Nations; 2019.
15. Waghulkar S, Ganjre K, Behare N, Diwan N. A feasibility study for online marketing of agricultural greenhouse products W.R.T. Pune District. Int J Manag. 2017;8(1):98–110.

Food Allergens and Related Computational Biology Approaches: A Requisite for a Healthy Life

Bhupender Singh, Arun Karnwal, Anurag Tripathi, and Atul Kumar Upadhyay

Introduction

Food allergy has become a global epidemic and the global population is experiencing food allergic conditions resulting in symptoms such as asthma, allergic rhinitis, etc. According to an estimate around 20% of the global population is under the influence of allergic conditions. In India, there is a dearth of detailed studies on allergy and most of the work in this field has been carried out by developed countries [33]. In developed countries approximately 4% to 7% children and 3% to 6% of adults are under the influence of allergic conditions. Unintentional introduction of the allergens to the allergenic individual may result in deadly anaphylactic response [14, 34]. Food allergy is termed as individual's inappropriate immune response against food. The mounted immune response as classified by NIAID (National Institute of Allergy and Infectious Diseases) can be triggered by IgE or without IgE or by both. IgE mounted immune response symptoms commonly arise in two hours after the allergenic food is consumed. Sensitisation process takes place prior to the production of IgE antibodies (specific to the food allergen) by plasma cells, which are differentiated, from B-lymphocytes. These antibodies get attached to the cell surface of mast cells and basophils and when individual gets exposed to the same allergen for second time, the antigenic part of the food allergen attaches to these

B. Singh · A. Karnwal
School of Bioengineering and Biosciences, Lovely Professional University, Jalandhar, Punjab, India

A. Tripathi
Food, Drug and Chemical Toxicology Group, CSIR-Indian Institute of Toxicology Research, Vishvigyan Bhawan, Lucknow, India

A. K. Upadhyay (✉)
Department of Biotechnology, Thapar Institute of Engineering & Technology, Patiala, India
e-mail: atul.upadhyay@thapar.edu

A. K. Upadhyay et al. (eds.), *Bioinformatics for agriculture: High-throughput approaches*, https://doi.org/10.1007/978-981-33-4791-5_9

antibodies resulting in release of histamine and leukotrienes. These chemicals are responsible for display of allergenic symptoms. Immune response triggered by immune cells comes under without-IgE triggered immune response. These responses lead to conditions such as enterocolitis, proctocolitis and enteropathy syndromes. These conditions are majorly found in children and infants showing symptoms of vomiting, diarrhoea, abdominal cramps, and sometimes blood along with faecal matter discharge. Blend of both IgE and without-IgE mediated immune response leads to conditions such as atopic dermatitis and eosinophilic esophagitis [42]. It can be summarised that after the entry of allergens into the body through skin, mucosa or gastrointestinal tract, food allergy is triggered by IgE immune response characterised by allergenic reactions like urticaria, abdominal pain, wheezing, etc. Release of specific set of chemicals (histamine, leukotriene) by immune system cells mediates these reactions.

Proteins present in food account for the allergenic reactions in susceptible individuals. In allergy-sensitive persons, these reactions may vary from mild allergic reactions to severe problems leading to death. Ninety percent of the food allergies have been reported from common foods like egg, shellfish, milk, fish, peanut, soy, nuts, tree and wheat. At present food industry identifies allergens present in food with the help of molecular biology techniques such as ELISA (Enzyme-Linked Immunosorbent Assay). However, due to cross-reactivity of proteins and complex food matrix, false-positive results of allergens are also reported. Last few years have witnessed a significant shift to overcome these false results by the use of MS (Mass Spectrometry) methods. MS-based methods have been clubbed with separation techniques such as liquid chromatography, which confirms its importance in food allergen detection without producing any bias in the results [31].

Extensive work of the several research groups across the world has resulted in the detail classification, pathogenesis, diagnosis and therapy for the food allergens and allergy mechanism. On the basis of the mechanism of the allergic reactions, adverse food reactions are classified into two groups, namely food intolerance and food allergy. Food intolerance referred to the toxic component, pharmacologic active part in the food or any physiological or the metabolic disorder in the host body which lead to the allergic conditions. Basically, it is food intolerance which accounts for majority of the adverse food reactions. Food allergy is defined as the atypical immune response to the food present in the host body. Further the food allergy is classified based on the type of the immune response triggering the food allergy. These categories are (a) immunoglobulin type E triggered, (b) immune cells triggered food allergy and (c) the food allergy triggered by involvement of immunoglobulin type E and immune cells. In IgE mediated allergenic responses, history plays a vital role as the allergic reactions take place right after the intake of the food resulting in the distress to numerous body organs. In allergic condition like enterocolitis and eosinophilic, esophagitis history does not play vital role as the symptoms appear after hours or days after the food intake.

The first and the foremost recommendation to patients with a food allergy is allergen avoidance. Therefore, the regulatory authorities of the government in countries like Japan and USA introduced the law to label the potential allergic food such as milk, eggs, peanuts, wheat, buckwheat, soybean, shrimp and crab, etc., to eliminate the intake of such food by the allergic patients. In another advancement in the field of food allergy cure Chinese herbal therapies, which are the combination of the unique herbs, were successful when tested on murine models of peanuts and anaphylaxis; however, the trials were limited to the mice only [8].

Identification of Food Allergens

As we have already discussed in the earlier sections of the review that there are several food allergens found in different foods. Here, in this section, a detailed discussion of the identification of various food allergens in many crops such as peanut, mustard, etc. by several molecular biology experiments and computational methods was performed. In order to identify the allergenic nature of proteins present in food against any individual under consideration prior knowledge of exposure to the food acts as essential factor and serves as a significant factor when individual's family also inherits the related food allergy.

Peanut allergy is a significant health problem, as it accounts for the anaphylactic reactions in peanut-sensitive individuals. Ara h 1 is an allergen produced in peanut, which was amplified in the crop with the use of specific primers of the allergen by PCR (Polymerase Chain Reaction) method. Northern blotting experiments have confirmed the presence of Ara h 1mRNA clone of 2.3-kilo base-pair (kb) in size. Sequencing and homology search approaches have established the homologous relationship of the Ara h 1 allergen with plant seed storage proteins. In another study, the Ara h 1 clone was expressed in *E. coli,* and its identification was carried out with help of serum of peanut-sensitive individual by immunoblotting technique [5].

Another essential edible oil crop is mustard, which also contributes to a significant proportion of food allergy. Braj1 from oriental mustard was separated and identified as an allergen. The allergen belongs to 2S albumin. The protein was identified as an allergen by immunoassays using IgE from mustard-allergenic individuals. This allergen was showing homology with yellow mustard seed allergen Sin a 1. In one the scientific study to validate the presence of Sin a 1, single precipitation band of all isoforms was seen in double diffusion immunoassay employing Sin a 1 targeted rabbit polyclonal serum [16].

Wheat allergens are categorised into water dissolving and non-dissolving proteins. There are several water dissolving allergens present in wheat such as profilin, agglutinin, thioredoxin, thiol reductase homologue, triosephosphate isomerase, 1-cys-peroxiredoxin, serpin, glyceraldehyde-3-phosphate-dehydrogenase, dehydrin, α-purothionin, protein resembling with inhibitor (serine protease), glutathione transferase, allergen resembling with thaumatin and peroxidase. Water non-dissolving allergens are α/β-gliadin protein, γ-gliadin protein, omega-1 protein, 2-gliadin protein, omega5-gliadin protein and glutenin subunits having elevated and lower molecular weight.

In chicken eggs, ovomucoid, ovalbumin, ovotransferrin and lysozyme are egg white allergens, whereas serum albumin and part of vitellogenin-1 precursor were egg yolk allergens [1, 24]. In cow's milk α-lactalbumin, immunoglobulin and casein are capable of inducing allergenic reactions in milk allergy susceptible individuals [10]. Allergenic proteins found in shrimp are triosephosphate isomerase allergenic protein, myosin light chain 2 allergenic proteins, arginine kinase allergenic protein, myosin light chain 1 allergenic protein, sarcoplasm calcium-binding allergenic protein, troponin C allergenic protein and tropomyosin allergenic protein [28].

A research was carried out which confirmed the activity of polyphenol oxidase (PPO) as an allergen in the eggplant (*Solanum melongena L.*). Various molecular biology techniques and computational methods were employed for the confirmation of polyphenol oxidase as an allergen. The eggplant proteins were separated from the peel extract on phenyl sepharose. The Enzyme-Linked Immunosorbent Assay and IgE Immunoblotting were used for the analysis of polyphenol oxidase activity. Mass spectrometry was used for the identification of the allergens and *in-silico* methodology was taken into account for the prediction of IgE competent epitopes. The eggplant allergens were classified into five components named as PS1, PS2, PS3, PS4 and PS5. PS2 showed off elevated polyphenol oxidase activity and this PS2 component was found in six eggplant-allergic individuals taken into consideration. It was reported that 43 kDa, 64 kDa and 71 kDa proteins showed sturdy competence with IgE. Two of them, viz., 64 kDa and 71 kDa proteins showed polyphenyl oxidase activity and were recognised as having polyphenol oxidase based on the existence of copper and detection by anti-sweet potato polyphenyl oxidase antiserum. The 43 kDa proteins were regarded as degraded component of 64 kDa or 71 kDa proteins based on its enzymatic activity and identification by polyphenol oxidase antiserum. They Further analysis the with the help of Mass Spectrometry helped in the identification of two more components named as PPO1 and PPO4. With the help of BCPRED [39], 15 overlapping B-cell epitopes in PPO4 were found. Computational prediction and analysis of PPO4 by AlgPred [38] resulted in the presence of IgE epitope **"PKPGMGTIEN"** (231–240 residues) and **"LKPGVDTIEN"** [18].

Impact of Food Processing on Allergenicity Potential of Food Allergens

Numerous research work has been conducted around the world in last several years to find the effect of treated food on their antigenic and allergenic potential. When milk was heat-treated the whey proteins get denatured gradually, while casein does not affect heat treatment because of the absence of any 2°, 3° and 4° protein structure, which leads to partial decrease in the allergenicity. Homogenisation of milk does not have any role in altering allergenic potential of the milk. Sterilisation also showed that only around 25% of whey proteins left intact while remaining portion undergoes denaturation and triggered Maillard reaction [4, 20, 29, 36] resulting in significant loss of allergic nature of milk and edible products made from milk. Several studies have been performed to check the effect of heat on the allergenicity of egg, which results in reduction of allergic symptoms in 50–85% of kids with egg allergy [9, 26, 41].The allergenicity of hazelnuts also reduced by significant amount on the consumption of baked products of it [17, 43].

Boiling of the peanuts in water at 100 °C for 20 min reduced immunoglobulin type E attaching efficiency for peanut allergenic proteins confirmed by immunoblotting. Autoclaving (2.56 atm for 30 min) of the roasted peanuts also decreased immunoglobulin type E attaching efficiency for peanut allergenic proteins. Hydrolysis effect on roasted peanut proteins was studied which results in hydrolysis of peroxidase, digestive enzymes and reduced level of allergens, viz., “Ara-h-1, Ara h 3” were observed while no effect was observed on raw peanut [2, 3, 6, 7, 44].

The sequence similarity among the allergenic protein from different sources has caused similar allergenic reactions, the process is known as cross-reactivity among allergens. Allergy causing protein sequence from peanut, which response to IgE of susceptible person’s sera, is having homology with the allergenic sequences of other foods such as soy, legumes and tree nuts. The allergen Jun a 3, from mountain cedar contains the similar protein sequence as of pepper, cherry, kiwi, tomato and apple.

The cross-reactivity among food and aeroallergens of animal, plant and fungal type has also been studied. It was reported that conditions of respiratory allergy patients having cross-reactivity of aero and food allergens might result in oral allergy syndromes, which may extend into severe anaphylaxis. The allergens which belong to various allergen families in cross-reactivity of food and aeroallergens are listed in tabular form (Table 1) [35]. Various syndrome and association, which arises by cross-reactivity of food and aeroallergen of plant type, are also listed in Table 2.

There are several syndromes associated with the cross-reactive food and aeroallergens few examples are: (a) birch-apple syndrome caused by Bet v 1 homologue Mal d 1 allergen, (b) Cypress-peach syndrome caused by Pru p 3 non-specific Lipid Transfer Protein, (c) Celery-mugwort-spice syndrome caused by Art v 4 profilin and Art v 60 kDa homologue to Api g 5 and (d) Mugwort-mustard syndrome may cause by Art v 3 LTP, Art v 4 profilin and Art v 60 kDa. Mugwort-peach association caused by Art v 4 profilin and Art v 3 LTP, Mugwort-chamomile association may cause by Art v 1 defensin, Ragweed-melon-banana association may be caused by Amb a 6 LTP and Amb a 8 profilin and Goosefoot-melon association may cause by Che a 2 profilin. Alternaria-spinach syndrome includes Alt a 1 allergen, Mite-shrimp syndrome includes Der p 10 tropomyosin, Cat-pork syndrome includes Fel d 2 cat serum albumin and Bird-egg syndrome includes Gal d 5 alpha-livetin which is chicken serum albumin allergen [35].

Table 1 Cross-reactivity of food and aeroallergens

Cross-reactive group	Allergen name	Source organism (common name)	Allergen classification (food/aero)
Number 1	Bet v 1	European White Birch	Aeroallergens
	Aln g 1	European Alder	
	Mal d 1	Apple	Food allergens
	Pru p 1	Peach	
	Api g 1	Celery	
	Gly m 4	Soybean	
Number 2	Bet v 2	European White Birch	Aeroallergens
	Ole e 2	Olive	
	Che a 2	Lambsquarters	
	Art v 4	Mugwort	
	Amb a 8	Short Ragweed	
	Api g 4	Celery	Food allergens
	Dau c 4	Carrot	
	Pru p 4	Peach	
	Cuc m 2	Muskmelon	
	Mus xp 1	Banana	
	Sin a 4	Yellow Mustard	
Number 3	Pla a 3	London Plane Tree	Aeroallergens
	Ole e 7	Olive	
	Art v 3	Mugwort	
	Amb a 6	Short Ragweed	
	Api g 2	Celery	Food allergens
	Pru p 3	Peach	
	Cuc m LTP	Musk Melon	
	Mus a 3	Banana	
	Sin a 3	Yellow Mustard	
Number 4	Der p 10	European House Dust mite	Aeroallergens
	Bla g 7	German Cockroach	
	Pen m 1	Black Tiger Shrimp	Food allergens
	Myt e 1	Mussel	
Number 5	Fel d 2	Cat	Aeroallergens
	Can f 3	Dog	
	Equ c 3	Domestic Horse	
	Bos d 6	Domestic Cattle	Food allergens
	Sus s 6	Domestic Pig	

Table 2 Various syndromes arising due to cross-reactive food and aeroallergens

Name of syndrome	Allergens involved (cross-reactive food and aeroallergens)
Birch-apple syndrome	Mal d 1(Apple) homolog to Bet v 1 (European White Birch)
Cypress-peach syndrome	Pru p 3 (Peach) (Non-specific lipid transfer protein)
Celery-Mugwort-Spice Syndrome	Art v 4 (Mugwort), Api g 5 (Celery) homolog to Art v 60 kDa
Mugwort-Peach association	Art v 4 (Mugwort) profilin, Art v 3 lipid transfer protein

Allergen Proteins in Various Pfam (Protein Domain) and Structural Families

A detailed analysis of available allergens and their classification among different protein families and motifs were performed in 2008. It was also suggested that a similar distinct region on allergen proteins having homologous structure might be of potential significance than whole sequence similarity. To find those regions on allergen proteins, these proteins were categorised in their subdomains present in SDAP (Structural Database of Allergenic Proteins) and into their respective protein family by the use of the Pfam database. Those SDAP allergenic proteins were categorised into 130 different Pfam families, out of which 31 families contain at least four allergens.

The PfamA domain family named Protease inhibitor/Lipid Transfer Protein family/seed storage with Pfam code PF00234 had 34 allergens. The mentioned domain was found in plants, and 3D structures of three allergen protein (Pru p 3, Hor v 1 and Zea m 14) were recognised. The t-cell epitope of allergen Ara h 2 belonging to this domain was also identified. The allergen belonging to this domain was named as Amb a 6 (short ragweed), Ara h 6 (peanut), Bra n 1(rapseed), Gly m 1(soybean), Hor v 21(barley), Lyc e 3(tomato), Par j 1 and Par j 2(*Parietaria judaica*), Pru av 3(sweet cherry), Pyr c 3(pear), Ses i 2(sesame), Tri a glutenin (wheat), Zea m 14(corn), Ana o 3(cashew nut), Ber e 1(Brazil nut), Cor a 8 (hazelnut), Hev b 12(rubber), Jug n 1(black walnut), Mal d 3(apple), Pru d 3(-European plum), Ric c 1(castor bean), Sin a 1(yellow mustard), Tri a TAI(wheat), Ara h 2(peanut), Bra j 1(oriental mustard), Fag e 8 kDa(common buckwheat), Hor v 1(barley), Jug r 1(English walnut), Ory s TAI(rice), Pru ar 3(apricot), Pru p 3(peach), Ses i 1(sesame), Tri a gliadin(wheat) and Vit v 1(grape). In the light of these studies, it was concluded that conserved sequence distinct from seed storage proteins, tropomyosin and Bet v 1 allergen family overlie with identified IgE epitopes which claim their motif-based advances to find allergenic potential of novel proteins [22]. Characterisation of Tryp_alpha_amyl protein family of plants was carried out to elucidate its relevance as food allergen protein. Out of the eight allergenic proteins in rice, seven allergens were categorised under Tryp_alpha_amyl protein family (trRSAs). The trRSAs genes were found on chromosome number 7, and their expression was dominant in fully grown seeds. Around 75 homologue proteins across 22 plant species which includes rice, maize, wheat and sorghum were reported for trRSAs protein family, collectively called as trHAs. These 75 homologue proteins were categorised into three groups named as Lipid Transfer Protein, seed storage protein and trypsin alpha amylase inhibitors. In another prominent work food allergen was analysed and classified into protein fold families on the basis of their structural configuration. The potential epitopes of these allergens were also mapped on the basis of structure.

Allergens were classified into several groups based on their structural families present in different structural databases such as SCOP (Structural Classification of Proteins) [19] and CATH (Class Architecture Topology Homologous superfamily) [12].One example is PR-10/Bet v 1 group which includes pollen allergen proteins, seeds allergen proteins and fruit tissue allergen proteins. Prolamin superfamily includes protease inhibitor allergenic protein, seed storage allergenic protein and lipid transfer allergenic protein families. Their characteristic is occurrence of conserved pattern of cysteine residues. Cupin superfamily presents retained, beta-barrel domain families. EF-hand family presents motifs which were Ca attaching sequence, primarily seen in parvalbumin. Twelve amino acid residues, forming ring like region were found to be preserved. Profilin family consists of plant profilins having 12–15 kDa vastly conserved proteins. Ara h 5 allergen comes under this family.

The relation between asthma and food allergy is well studied and established by several independent efforts. These two conditions are most common in many children and can lead to severe effects on the health of the patient. Children developing these two conditions are prone to anaphylaxis, which can be mortal. The number of patients of asthma and food allergy are developing simultaneously in number in past decades.

Computational Methods to Predict Food Allergens

There are several useful resources and databases for detailed information about allergens such as AllAllergy, International Union of Immunological societies, Allergome [27], Central Science Laboratory UK, National Center of Food Safety and Allergy, Protall, InformAll, ADFS (Allergen Database for Food Safety) [32], Allermatch [15], WebAllergen [37], Algpred and SDAP (Structural Database of Allergenic Proteins) [21].

SDAP provides the efficacy of query protein match to the allergen. This database was brought into picture the character of allergen proteins and for the peer researchers to check whether the test protein has allergenic nature or not. They have also mentioned the way to check the cross-reactivity and methods by which allergenic potential can be predicted. The listed methods are WebAllergen, Allermatch and AlgPred out of which on predicting only WebAllergen was able to discriminate between tropomyosin and allergenic tropomyocin by presence of extra allergenic motifs. They also mentioned about PCPMer suite which performs motif search for checking the allergenic potential [40].

Allergen prediction tool named as AlgPred follows the *in-silico* approach to predict the potential allergens in the given query protein sequence. It also finds their respective immunoglobulin type E antigens if present. Features like peptide residues, dipeptide constitution retained 85.02%, 84% precision and specificity, respectively. A web-based tool named as AllerTool for finding the allergenic potential and cross-reactivity among the allergen proteins was also developed. This tool also provides the machine learning (support vector machine) based approach to find the allergenic potential with 86% sensitivity and specificity, respectively [23]. Machine learning approach was implemented to find out the allergenic proteins by means of evolutionary relationship and presented the results, which claims increased sensitivity and specificity. In SVM, features like protein-peptide residues, dipeptide construction, simulated amino acid construction and PSSM were used. For the validation of the Support Vector Machine model, 10-fold cross-validation method was used in which dataset was arbitrarily categorised into 10 subsets. The accuracy of the SVM model which uses PSSM was also compared with existing allergen prediction models like Algpred and WebAllergen. They concluded that use of the evolutionary information could be of greater importance for developing the more accurate allergen prediction model [25]. Another work predicts the allergen proteins by the Chou's simulated amino acid construction and machine learning approaches. Chou's simulated amino acid construction methodology was developed which improves the allergen prediction efficiency of proteins located at subcellular location and membrane proteins. For the prediction, they also took use of machine learning approach into account. They used Support Vector Machine for the prediction, which considers vector presentation of the sequences obtained from sequence characteristics. In Chou's simulated amino acid construction, they took into account various characteristics of the amino acids, which were hydrophobic, hydrophilic, isoelectric point, pK1 and pK2 nature of the amino acids. Out of these properties they made a sum of 41 arrangements of the allergens and non-allergens for testing the datasets. In total the accuracy obtained on the datasets by them was 91.9%, and Mathew's correlation coefficient value obtained was o.82 [30]. ProAP, a tool is able to predict the allergenic nature of the query protein by worldwide search and stated as influential tool to predict the allergens. The web server-based tool was available at http://gmobl.sjtu.cn/proAP/main.html [45]. A tool named Allerdictor is used for the prediction of allergenic protein by the implementation of text classification method. The Allerdictor examined approximately 540,000 protein sequences from Swiss-Prot in around six minutes and identified less than 1% of sequences as allergenic [11]. Another significant development in the field was based on artificial neural network to check the allergenic potential of the various allergens. Two distinct algorithms were coded consisting of three and four steps, and their prediction potential was determined on 2427 positive entries and 2427 negative entries. The positive protein sequences retrieval sources—Central Science Laboratory allergen database, Food Allergen Research and Resource Program allergen database, SDAP and Allergome database while the non-allergens protein sequences of commonly used food such as bread wheat, tomato, potato, pepper, Asian and African rice were obtained from Swiss-Prot. In three-step algorithm, firstly physio-chemical characters of amino acids explained protein sequence, which includes size, hydrophobicity, and relative abundance and beta-strand and helix propensities. Secondly, they converted strings into vectors of same length via auto and cross-covariance. In last step, Artificial Neural Network was employed to distinguish the allergens from non-allergens. After this first distinguished web server-based tool named as AllerTop for the allergen prediction was introduced. This method presents the first non-alignment-based approach for the prediction. When the AllerTop was compared with other existing prediction servers, the AllerTop web server does better than all by giving 94% sensitivity. The AllerTop was made freely available at http://pharmfac.net/allertop [13]. Recently, a new algorithm named as PREAL^{W}was introduced, which combines PREAL, FAO/WHO methodology for allergen prediction and motif-based method and takes the average of the weighted score for the prediction. This method was termed as integrative because of the combination of various prediction methods in order to embed the characteristic of various methods and overcome the limitation of these predictions. This method was regarded as best suitable for the prediction of crops allergens (Table 3).

Conclusion

In our day to day life we come across several news of people showing allergy to different foods and with severe effect on their health. Sometime these allergies are so severe that it may cause death of the individuals. The immunoglobulin E (IgE) immune response is the primary reason for the triggering of food allergy, which is activated by the allergens entering the body. In case of allergic reactions, histamine and leukotriene are released by the activated mast cells and basophils. This chemical release leads to allergic responses such as itching, abdominal pain, dyspnea and wheezing. For our surprise, the majority of the commonly used food items have several allergens. For example, chicken eggs have ovomucoid, ovalbumin, ovotransferrin and lysozyme, which have an allergic effect on humans. In cow's milk α-lactalbumin, β-lactalbumin, casein (α_s1, αs2, β, κ) were the proteins found to be allergic. Several experimental, bioinformatics and computational studies have been conducted to gain the knowledge of IgE motif epitopes in allergens and transgenic proteins introduced in the genetically modified food to limit the use of allergen transgene in genetically modified food. In present study, we have performed a detailed literature survey of the studies related to food allergens. In many studies, computational approaches (machine learning approaches) such as Support Vector Machines and Random Forest classifiers were used to predict the common epitopes present on human proteins, which react, with these allergens to cause allergic reactions. There is an urgent need to analyse these epitopes and protein structures in detail to suggest critical residues for engineering purposes to avoid their life-threatening allergic effects.

In this review work we have started with general introduction to the food allergens followed by a detailed shedding of light on the prevalence of these food

Table 3 List of allergen prediction tolls and their important features

S. No.	Prediction tool	Features employed	Dataset	Remarks
1	AlgPred (14)	Amino acid composition, Dipeptide composition. Prediction by: (a) IgE epitope presence (b) Motif based and (c) ARP (Antigen Representing Peptides) based.	Obtained from http://www.slv.se/templatesSLV/SLV_Page_9343.asp 578 Allergens and 700 non-allergens.	For the first time SVM technique of machine learning has been implemented in allergen prediction with high accuracy.
2	AllerTool (44)	XR-BLAST, XR-Graph, ALR-SCAN and ALR-SVM.	For ALR-SVM same dataset was used as in AlgPred. For others, data obtained from Allergen.org and ALLERDB database. 373 allergens, 260 iso-allergens and 128 cross-reactivity cases.	First tool to anticipate the cross-reactivity among food allergen proteins.
3	SVM method based on evolutionary model. (45)	Amino acid composition, Dipeptide composition, Pseudo amino acid composition, PSSM (Position Specific Scoring Matrix).	Obtained from SDAP (Structural Database of Allergenic Proteins) and UniProt. 693 allergens and 1041 non-allergens.	Prediction results obtained by implementation of evolutionary information via PSSM in SVM outperform other SVM features (amino acid composition, dipeptide composition and pseudo amino acid composition) used in allergen prediction.
4	Prediction of IgE motif epitopes in proteins from genetically modified foods. (46)	Amino acid composition and Dipeptide composition.	Dataset obtained from NCBI (National Center for Biotechnology Information) and allergen database.	A dedicated resource developed to predict allergenic proteins in genetically modified food for developing food safety guidelines and immunotherapy strategies.
5	Allergenic protein prediction by Chou's pseudo amino acid	Pseudo amino acid composition.	Dataset obtained from: http://www.imtech.res.in/raghava/algpred.	Developed model was suggested to be most appropriate for predicting

(continued)

Table 3 (continued)

S. No.	Prediction tool	Features employed	Dataset	Remarks
	composition and machine learning. (47)		Contains 460 allergens and 560 non-allergens.	novel allergenic proteins.
6	ProAP (48)	Amino acid composition, Sequence based prediction and motif-based prediction.	Dataset curated from Swiss-Prot allergen index, IUIS allergen nomenclature, SDAP and ADFS. 989 allergens and 244,538 non-allergens were collected.	Provides personalised search criteria to predict allergenic proteins at ease.
7	Allerdictor (49)	Text classification technique coupled in SVM.	Dataset obtained from IUIS allergen nomenclature, Allergome, SDAP and AllergenOnline and AllerMatch. 3907 allergens and 464,101 non-allergens were used.	Generates fast allergen prediction results with almost equivalent/better accuracy with other existing methods. Suitable for prediction of allergen proteins in genetically modified crops and in large data.
8	Allergenicity prediction by Artificial Neural Networks. (50)	E-descriptors.	Dataset derived from the Central Science Laboratory (CSL) allergen database, FARRP, SDAP and Allergome database. 2427 allergens and non-allergens were used in study.	Provides universal ANN algorithm for allergen prediction. As some existing methods were efficient in allergen prediction and some in non-allergen, use of multiple predictors for allergen prediction was suggested.
9	AllerTop (51)	Z-descriptors.	Dataset curated from CSL allergen database, FARRP and SDAP. 2210 allergens and non-allergens were used.	First alignment-independent approach to predict the allergen proteins.
10	PREAL^{w} (52)	Amino acid composition, physico-chemical properties (Implemented in PREAL tool),	Dataset curated from WHO/IUIS allergen nomenclature, Allergome, FARRP Allergen	Combinations of approaches were used for allergen prediction.

(continued)

Table 3 (continued)

S. No.	Prediction tool	Features employed	Dataset	Remarks
		motif-based prediction and FAO/WHO criteria for allergen prediction.	Online and Allfam. 830 allergens and 298,827 non-allergens were used	Significant for allergen prediction in crops.

allergens in different protein domain families and folds. We have reviewed the tools and databases used in the identification and analysis of different food allergens. We have also presented our views and suggestion along with current scenario regarding the development of the research opportunities and scopes in the field of food allergy.

Ethics Approval and Consent to Participate
Not applicable.

Consent for Publication
Not applicable.

Availability of Data and Materials
The authors have agreed to provide the data and documents for open access.

Competing Interests
The authors declare that they have no competing interests.

Funding
Not Applicable.

References

1. Amo A, Rodríguez-Pérez R, Blanco J, Villota J, Juste S, Moneo I, Caballero ML. Gal d 6 is the second allergen characterized from egg yolk. J Agric Food Chem. 2010;58(12):7453–7. https://doi.org/10.1021/jf101403h.
2. Baumert JL, Verhoeckx KCM, Flanagan S, Herouet-Guicheney C, Shimojo R, van der Bolt N, Vissers YM, et al. Food processing and allergenicity. Food Chem Toxicol. 2015;80:223–40. https://doi.org/10.1016/j.fct.2015.03.005.
3. Beyer K, Morrow E, Li XM, Bardina L, Bannon GA, Wesley Burks A, Sampson HA. Effects of cooking methods on peanut allergenicity. J Allergy Clin Immunol. 2001;107(6):1077–81. https://doi.org/10.1067/mai.2001.115480.
4. Bu G, Luo Y, Chen F, Liu K, Zhu T. Milk processing as a tool to reduce cow's milk allergenicity: a mini-review. Dairy Sci Technol. 2013;93(3):211–23. https://doi.org/10.1007/s13594-013-0113-x.
5. Burks AW, Cockrell G, Steven Stanley J, Helm RM, Bannon GA. Recombinant peanut allergen Ara h I expression and IgE binding in patients with peanut hypersensitivity. J Clin Investig. 1995;96(4):1715–21. https://doi.org/10.1172/JCI118216.
6. Cabanillas B, Maleki SJ, Rodríguez J, Burbano C, Muzquiz M, Jiménez MA, Pedrosa MM, Cuadrado C, Crespo JF. Heat and pressure treatments effects on peanut allergenicity. Food Chem. 2012;132(1):360–6. https://doi.org/10.1016/j.foodchem.2011.10.093.

7. Chung SY, Maleki SJ, Champagne ET. Allergenic properties of roasted peanut allergens may be reduced by peroxidase. J Agric Food Chem. 2004;52(14):4541–5. https://doi.org/10.1021/jf030808d.
8. Cianferoni A, Spergel JM. Food allergy: review, classification and diagnosis. Allergol Int. 2009;58(4):457–66. https://doi.org/10.2332/allergolint.09-rai-0138.
9. Cortot CF, Sheehan WJ, Permaul P, Friedlander JL, Baxi SN, Gaffin JM, Dioun AF, Hoffman EB, Schneider LC, Phipatanakul W. Role of specific IgE and skin-prick testing in predicting food challenge results to baked egg. Allergy Asthma Proc. 2012;33(3):275–81. https://doi.org/10.2500/aap.2012.33.3544.
10. D'Urbano LE, Pellegrino K, Artesani MC, Donnanno S, Luciano R, Riccardi C, Tozzi AE, Ravà L, De Benedetti F, Cavagni G. Performance of a component-based allergen-microarray in the diagnosis of cow's milk and hen's egg allergy. Clin Exp Allergy. 2010;40(10):1561–70. https://doi.org/10.1111/j.1365-2222.2010.03568.x.
11. Dang HX, Lawrence CB. Allerdictor: fast allergen prediction using text classification techniques. Bioinformatics. 2014;30(8):1120–8. https://doi.org/10.1093/bioinformatics/btu004.
12. Dawson NL, Lewis TE, Das S, Lees JG, Lee D, Ashford P, Orengo CA, Sillitoe I. CATH: an expanded resource to predict protein function through structure and sequence. Nucleic Acids Res. 2017;45(D1):D289–95. https://doi.org/10.1093/nar/gkw1098.
13. Dimitrov I, Bangov I, Flower DR, Doytchinova I. AllerTOP v.2 - a server for in silico prediction of allergens. J Mol Model. 2014;20(6) https://doi.org/10.1007/s00894-014-2278-5.
14. Dunlop JH, Keet CA. Epidemiology of food allergy. Immunol Allergy Clin North Am. 2018;38(1):13–25. https://doi.org/10.1016/j.iac.2017.09.002.
15. Fiers MWEJ, Kleter GA, Nijland H, Peijnenburg AACM, Nap JP, van Ham RCHJ. Allermatch^{TM}, a Webtool for the prediction of potential allergenicity according to current FAO/WHO Codex Alimentarius Guidelines. BMC Bioinformatics. 2004;5:1–6. https://doi.org/10.1186/1471-2105-5-133.
16. González de la Peña MA, Menéndez-Arias L, Monsalve RI, Rodríguez R. Isolation and characterization of a major allergen from oriental mustard seeds, BrajI. Int Arch Allergy Appl Immunol. 1991;96(3):263–70. https://doi.org/10.1159/000235505
17. Hansen KS, Ballmer-Weber BK, Lüttkopf D, Skov PS, Wüthrich B, Bindslev-Jensen C, Vieths S, Poulsen LK. Roasted hazelnuts - allergenic activity evaluated by double-blind, placebo-controlled food challenge. Allergy: Eur J Allergy Clin Immunol. 2003;58(2):132–8. https://doi.org/10.1034/j.1398-9995.2003.23959.x.
18. Harish Babu BN, Wilfred A, Venkatesh YP. Emerging food allergens: identification of polyphenol oxidase as an important allergen in eggplant (Solanum Melongena L.). Immunobiology. 2017;222(2):155–63. https://doi.org/10.1016/j.imbio.2016.10.009.
19. Hubbard TJP, Murzin AG, Brenner SE, Chothia C. SCOP: A Structural Classification of Proteins Database. Nucleic Acids Res. 1997;25(1):236–9. https://doi.org/10.1093/nar/25.1.236.
20. Huffman LM, De Barros Ferreira L. Dairy ingredients for food processing; 2011. https://doi.org/10.1002/9780470959169.ch8.
21. Ivanciuc O, Schein CH, Braun W. SDAP: database and computational tools for allergenic proteins. Nucleic Acids Res. 2003;31(1):359–62. https://doi.org/10.1093/nar/gkg010.
22. Ivanciuc O, Torres M, Braun W, Garcia T, Schein CH. Characteristic motifs for families of allergenic proteins. Mol Immunol. 2008;46(4):559–68. https://doi.org/10.1016/j.molimm.2008.07.034.
23. Koh JLY, Zhang ZH, Tong JC, Zhang GL, Choo KH, Tammi MT. AllerTool: a web server for predicting allergenicity and allergic cross-reactivity in proteins. Bioinformatics. 2006;23(4):504–6. https://doi.org/10.1093/bioinformatics/btl621.
24. Kondo Y, Tsuge I, Tanaka A, Sampson HA, Sicherer SH, Shreffler WG, Noone S, et al. Original Article Chicken Serum Albumin (Gal d 5 *) is a partially heat-labile inhalant and food allergen implicated in the bird-egg syndrome. Allergol Int. 2013;58(5):481–6. https://doi.org/10.2332/allergolint.12-OA-0513.

25. Kumar KK, Shelokar PS. An SVM method using evolutionary information for the identification of allergenic proteins. Bioinformation. 2012;2(6):253–6. https://doi.org/10.6026/97320630002253.
26. Lemon-Mulé H, Sampson HA, Sicherer SH, Shreffler WG, Noone S, Nowak-Wegrzyn A. Immunologic changes in children with egg allergy ingesting extensively heated egg. J Allergy Clin Immunol. 2008;122(5):977–84. https://doi.org/10.1016/j.jaci.2008.09.007.
27. Mari A, Rasi C, Palazzo P, Scala E. Allergen databases: current status and perspectives. Curr Allergy Asthma Rep. 2009;9(5):376–83. https://doi.org/10.1007/s11882-009-0055-9.
28. Matsuo H, Yokooji T, Taogoshi T. Common food allergens and their IgE-binding epitopes. Allergol Int. 2015;64(4):332–43. https://doi.org/10.1016/j.alit.2015.06.009.
29. Michalski MC, Januel C. Does homogenization affect the human health properties of cow's milk? Trends Food Sci Technol. 2006;17(8):423–37. https://doi.org/10.1016/j.tifs.2006.02.004.
30. Mohabatkara H, Beigib MM, Abdolahic K, Mohsenzadeh S. Prediction of allergenic proteins by means of the concept of Chou's pseudo amino acid composition and a machine learning approach. Med Chem. 2013;9(1):133–7. https://doi.org/10.2174/157340641130901013.
31. Monaci L, De Angelis E, Montemurro N, Pilolli R. Comprehensive overview and recent advances in proteomics MS based methods for food allergens analysis. TrAC - Trends Anal Chem. 2018;106:21–36. https://doi.org/10.1016/j.trac.2018.06.016.
32. Nakamura R, Teshima R, Takagi K, Sawada J. Development of Allergen Database for Food Safety (ADFS): an integrated database to search allergens and predict allergenicity. Kokuritsu Iyakuhin Shokuhin Eisei Kenkyujo Hokoku= Bulletin of National Institute of Health Sciences. 2005;123:32–6.
33. Nanda MS, Singh K, Devi R. Allergy profile of patients visiting a tertiary care hospital in hilly areas of Solan, Himachal Pradesh, India. J Clin Diagn Res. 2018;12(2):MC01–3. https://doi.org/10.7860/JCDR/2018/34780.11161.
34. Nwaru BI, Hickstein L, Panesar SS, Muraro A, Werfel T, Cardona V, Dubois AEJ, et al. The epidemiology of food allergy in Europe: a systematic review and meta-analysis. Allergy: Eur J Allergy Clin Immunol. 2014;69(1):62–75. https://doi.org/10.1111/all.12305.
35. Popescu F-D. Cross-reactivity between aeroallergens and food allergens. World J Methodol. 2017;5(2):31. https://doi.org/10.5662/wjm.v5.i2.31.
36. Porter JWG. Discussed with the possibility of Maillard reaction. The future developments in milk protein with its high nutritional value and the necessity for new ways of using it as a food are stressed. J Soc Dairy Technol. 1978;31(4):199–202.
37. Riaz T, Hor HL, Krishnan A, Tang F, Li KB. WebAllergen: a web server for predicting allergenic proteins. Bioinformatics. 2005;21(10):2570–1. https://doi.org/10.1093/bioinformatics/bti356.
38. Saha S, Raghava GPS. AlgPred: prediction of allergenic proteins and mapping of IgE epitopes. Nucleic Acids Res. 2006;34(WEB. SERV. ISS):202–9. https://doi.org/10.1093/nar/gkl343.
39. Saha S, Raghava GPS. BcePred: prediction of continuous B-cell epitopes in antigenic sequences using physico-chemical properties; 2010. p. 197–204. https://doi.org/10.1007/978-3-540-30220-9_16.
40. Schein CH, Ivanciuc O, Braun W. Bioinformatics approaches to classifying allergens and predicting cross-reactivity. Immunol Allergy Clin North Am. 2007;27(1):1–27. https://doi.org/10.1016/j.iac.2006.11.005.
41. Turner PJ, Mehr S, Joshi P, Tan J, Wong M, Kakakios A, Campbell DE. Safety of food challenges to extensively heated egg in egg-allergic children: a prospective cohort study. Pediatr Allergy Immunol. 2013;24(5):450–5. https://doi.org/10.1111/pai.12093.
42. Wood R, Tang M, Lack G, Ebisawa M, Sicherer S, Eigenmann PA, Chiang W, et al. ICON: food allergy. J Allergy Clin Immunol. 2012;129(4):906–20. https://doi.org/10.1016/j.jaci.2012.02.001.
43. Worm M, Hompes S, Fiedler EM, Illner AK, Zuberbier T, Vieths S. Impact of native, heat-processed and encapsulated hazelnuts on the allergic response in hazelnut-allergic patients. Clin Exp Allergy. 2009;39(1):159–66. https://doi.org/10.1111/j.1365-2222.2008.03143.x.

44. Yu J, Ahmedna M, Goktepe I, Cheng H, Maleki S. Enzymatic treatment of peanut kernels to reduce allergen levels. Food Chem. 2011;127(3):1014–22. https://doi.org/10.1016/j.foodchem.2011.01.074.
45. Zhang D, Li J, Yu Y, Wang J, Zhao Y. Evaluation and integration of existing methods for computational prediction of allergens. BMC Bioinformatics. 2013;14(Suppl 4):S1. https://doi.org/10.1186/1471-2105-14-s4-s1.

GPSR Compliance

The European Union's (EU) General Product Safety Regulation (GPSR) is a set of rules that requires consumer products to be safe and our obligations to ensure this.

If you have any concerns about our products, you can contact us on ProductSafety@springernature.com

In case Publisher is established outside the EU, the EU authorized representative is:

Springer Nature Customer Service Center GmbH
Europaplatz 3
69115 Heidelberg, Germany

Batch number: 10264717

Printed by Printforce, the Netherlands